PAPER
MAKING
AND BOOKBINDING
Coastal Inspirations

PAPER MAKING

AND BOOKBINDING

Coastal Inspirations

Joanne B. Kaar

GUILD OF MASTER CRAFTSMAN PUBLICATIONS

Acknowledgements

I would like to thank the following for all their advice and help, which enabled me to write this book: Joe Kaar; Michael O'Donnell; Liz O'Donnell; Cullen O'Donnell; Yosephin Mira Padmasari; Alice Hui Fang Lee; Pat Spark; Inge Evers; Peter Leadbetter, Kemtex Educational Supplies; John Wojciechoski, The Scottish Bamboo Nursery; Nancy Ballesteros, Treetops Colours and Jeremy Youngs, John Purcell Paper.

First published 2003 by
Guild of Master Craftsman Publications Ltd
166 High Street, Lewes
East Sussex, BN7 1XN

Text © Joanne B. Kaar 2003
© in the work GMC Publications Ltd

ISBN 1 86108 286 X

British Cataloguing in Publication Data
A catalogue record of this book is available from the British Library.

Publisher: Paul Richardson
Art Director: Ian Smith
Production Manager: Matt Weyland
Managing Editor: April McCroskie
Editor: Gill Parris
Designed by Phil and Traci Morash at Fineline Studios

Photographic Credits
All photographs taken by Joanne B. Kaar, apart from step-by-step photographs showing hands, which were taken by Michael O'Donnell, and the picture of the author on the front French flap, which was taken by Joe Kaar.

Typeface: The Sans

Colour reproduction by Universal Graphics Pte Ltd, Singapore

Printed and bound in Hong Kong by CT Printing Ltd

Contents

Introduction 1

Silkpaper making 61

Section One:
PAPERMAKING

Colouring Silk Fibre 69

Section Two:
THE PROJECTS

Choosing Fibre for Papermaking 7

Types of Papermaking Frames 13

Decorative Envelopes 77

Is it Paper? 17

Japanese Stab Binding 99

Basic Materials and Equipment 21

Single-section Binding 129

How to Make a Mould and Deckle 27

Multi-section Binding 153

Glossary 180

Basic Pulp Papermaking Techniques 35

Useful Addresses 181

Colouring Pulp 49

Further Reading 182

Index 183

Introduction

Handmade papers have a special quality
all their own. Whether delicate and subtle
or robust and vibrant, no two sheets are the
same, even when made from the same vat
of pulp and by the same hands. The process
of making a sheet of paper is undeniably
fun, but it is the excitement of
experimenting with new decorative
techniques and colour combinations that
continually fascinates me.

A piece of music composed for
us by Jenny Broughton on the
occasion of moving into our
new house

I printed some of my gran's favourite recipes on handmade papers before binding them together

If inspiration evades you, start with your surroundings. I live on the far north coast of Scotland, and the sea and all it casts ashore are an ever-changing source of inspiration: driftwood with its sandblasted, sea-weathered textures; delicate patterns and bold shapes found in countless types of shells; mysterious colours floating in rock pools; brilliantly painted fishing boats. They are all a feast for the eye.

My grandad's feathers for fly fishing

Memories of our childhood

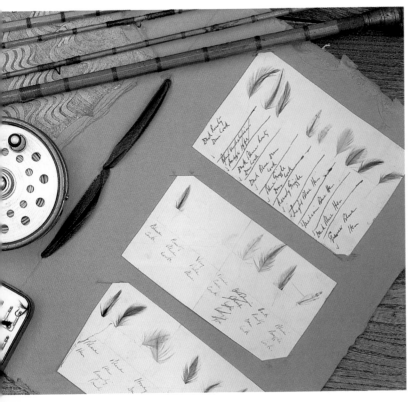

Here, I have combined these textures, forms and colours with a vast array of decorative techniques – from water cuts and rain paper, to printmaking and burning – to create a series of stunning handmade papers.

I show how the binding of a book can be creative, too: the stitching can be decorative as well as strong, and the threads and ribbons need not be traditional bookbinding materials. By using an imaginative approach to every aspect of binding, you will add a very special, personal touch.

With the projects shown here, my aim is to inspire you to make your own unique books, whether for cherished possessions, as a permanent reminder of a seaside holiday, a favourite walk, childhood memories or family recipes.

Threads, ribbons, tapes and string

Hemp
raw

Blue denim
raw loose fibres

Philippine
gampi

Kozo
raw

Wheat straw
raw

Cotton
raw

Flax
raw

Choosing fibre for papermaking

Raw plant fibre

Paper is made from cellulose fibres that form the tubular inner cell walls used to transport water through plants. This fibre can be released from plants by retting (rotting of fibre) and cooking in caustic soda, a strong alkali solution which is also used as a drain cleaner. These methods will remove the unwanted outer layer or bark of plants and the inner woody pith which contains lignin. If the lignin, which repels water, is left in the paper, the cellulose fibre will not be able to form a strong bond, so the resulting paper will be weak.

Once the cellulose fibre has been removed from the plant, it is then beaten to separate the filaments, a process called macerating. This process keeps the fibres long and also gives them a larger surface area. These are important qualities in making strong paper. If an alkali has been used, it must be rinsed out of the fibre. It is then ready for making paper.

Left Raw plants for papermaking

I keep my samples of plant papers made this way in a sketch-book, labelled for future reference

Papers made from plant fibre only

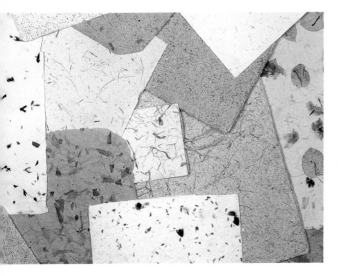

Papers made using wastepaper and plants

Retting fibre can take months. Beating fibre will take hours. Boiling fibre in caustic soda is dangerous, as it can burn your skin. But the rewards are great. Papers made from plants using these methods can look very beautiful, will be very strong, and they become very precious because of the effort making them.

Recycled paper

In a domestic situation, recycling wastepaper in a kitchen blender is much easier than starting with raw plant fibre, as the plant fibre has been processed for you. However, recycled paper may already have been been processed more than once, and you are about to process it again. Each time the fibres are processed they get shorter and, the shorter the fibre, the weaker the paper you make with it will be.

There will also be unknown chemicals in the wastepaper, so any new papers made from the waste will have an uncertain life expectancy and the printing inks may cause your paper to yellow with age. Some kinds of typing paper, photocopy paper, papers used for drawing and offcuts of mount board are all suitable. But, to avoid disappointment when planning a big project, experiment to find out which types of wastepaper can be recycled and what the limitations are of using wastepaper on its own. Choose a fibre that will make strong paper but will not damage your blender.

Handmade book with pages of wastepaper and plants. The stitching on the spine is coming apart, because the recycled wastepaper on its own is not strong enough for multi-sectioned binding

Be careful when choosing papers:

- First remove all items that may damage the blades on your blender, such as staples, paperclips, sticky tape and plastic windows found in business envelopes.
- The brilliant colours found in tissue paper and crêpe papers may look tempting but, if twisted into long strands, they can get tangled around the blender blades.

- Some papers, such as glossy magazines and gift wrapping, may not break down into a pulp because of the large amounts of chemicals and, often, plastic coatings.
- Although the high amounts of printing ink in newspapers may be removed with boiling, the paper itself is of a very poor quality.

All the projects in this book can be made with wastepaper, but the finished results may not be as strong as using alternative fibres, so experiment, following the instructions for basic pulp papermaking (see page 35).

Part-processed fibres linters, half-stuff and wet lap

Plant fibres that have been part-processed for use in papermaking are called half-stuff and wet lap, and these are sold in concentrated sheet form. When processing cotton fibre to make thread, it is passed through a linting machine. The waste cotton fibres left over from this process are then processed for papermaking. They are also sold in a concentrated sheet form and called linters. This makes half-stuff, wet lap and linters as easy to use as recycled waste paper, but they have the added strength of papers made from raw plant fibre, because the fibres have not been used before. There are many different plants that can be used to make paper, as an alternative

Linters, half-stuff and wet lap

to trees. They will make papers of varying strength, because each plant has different fibre lengths and some stem or bast fibres may be longer than leaf fibres. Before embarking on a big project, experiment with the different types available to make sure that you choose the right fibre for the job. I have found that bamboo, abaca, flax and hemp half-stuff, processed in a kitchen blender, make the strongest papers suitable for bookbinding. They can be used separately, or combined, depending on the qualities that you would like your handmade paper to have.

Bashania fargesii

As bamboo is the fastest-growing plant in the world, and is a farmed renewable crop, I have chosen to use the bamboo half-stuff for the projects in this book. This is sold by weight or by sheet, depending on where you purchase it. Usually 250g is the smallest amount for purchase if bought by weight.

Although bamboo half-stuff is easy to use and will make paper strong enough to bind into books, it has no character. It is just plain white pulp. A blank canvas. Through the projects in this book I will show you a variety of decorative techniques that will enable you to make distinctive, vibrant and individual papers with dyed bamboo half-stuff.

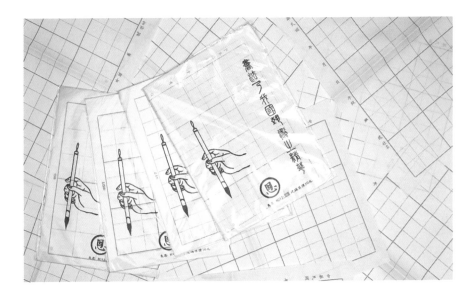

Calligraphy paper from Taiwan

Look at handmade books from around the world for inspiration

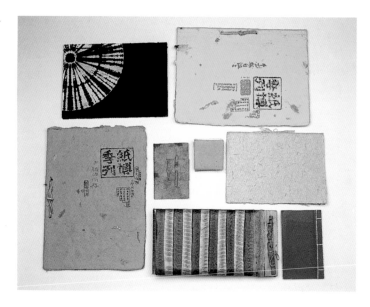

There are hundreds of species of bamboo, but the one used for production of this half-stuff is *Dendrocalamus strictus*, which grows up to 15m (48ft) and was harvested in Thailand as a farmed forestry product for paper. Although this is a subtropical species of bamboo, I have also found one that is used for making paper and will grow in Scotland, *Bashania fargesii*. It will only grow to 4m (12ft).

In China they have used strips of bamboo to write on since 500BC and have used bamboo for making paper since the eleventh century. It is now used in China to make paper in everyday use such as a practice paper for calligraphy in primary schools, and as 'spirit money', a ceremonial paper for burning, because it is a cheap source of strong fibre.

Ceremonial paper for burning, from Taiwan

Types of papermaking frames

Apprentice sugeta

There are many different frames for making paper by hand and I have chosen three basic types to compare. These frames can all be purchased (see 'Useful Addresses' on page 181) or, if you prefer, you can make your own (instructions are given on pages 27–33). Each one has individual qualities to take into consideration when deciding on the type of fibre you are using and the number of sheets of paper you want to make.

For all the projects in this book I have used the western style of mould and deckle for making papers, with bamboo half-stuff, because it is a simple technique to learn and allows for the quick and easy production of a large number of sheets of paper.

Sugeta, used for Oriental/Eastern papermaking

Left Sugeta

The mesh of a sugeta is made from strips of bamboo that have been stitched together to form a flexible screen. This flexible screen is called the su. It is trapped in a hinged wooden frame called a keta. When used together they are called a sugeta. Because the su is expensive to make, a rigid metal mesh is used by apprentice papermakers. The sugeta is suspended above the vat of pulp using a pulley system of ropes tied to the handles on the keta, making it possible for one person to make very large sheets of paper.

To slow down the drainage of water through the su and help separate the long fibres, a substance is added to thicken the water. This is called a neri, or formation aid, and can be made from many plants, including okra pods. Synthetic equivalents are available from suppliers of hand-papermaking materials.

By using a repeated wave action when dipping the sugeta in the vat, the excess pulp and water are thrown off the top, and paper is formed by a number of thin layers of long plant fibres gradually being built up. It is this repeated layering of long plant fibres that gives the paper its strength. The sugeta is left suspended over the vat of pulp while the su is removed and the freshly made paper is couched onto a stack of paper called a post, ready to be pressed. No felts or cloths are used, but single threads are placed between the sheets of paper. After pressing, but while still wet, the thin sheets are separated using the single threads, and stuck to a board for drying. To calculate the papermaker's wages, the threads are counted.

A sugeta can be constructed for use in the home by making two simple A4 frames the same as the deckle illustrated below left. Hinge the frames together to make the keta, then place a reed or bamboo placemat (which can be purchased from any kitchen shop) between the hinged frames to form the su.

Deckle Box and Mould, Used for Nepalese and Indian Papermaking

In Nepal and India papers are made using just a mould, which is a simple wooden frame with a fixed mesh, often made of cotton cloth. This is floated cloth-side down either on water in a container, or in a moving stream. Enough pulp for one sheet is poured into the floating mould, and fibres are agitated and distributed evenly, either by hand, or by the movement of the water in the stream. The mould is then lifted out of the water with the new sheet of paper formed inside it. It is then propped up in the sun and the paper left to dry while still in the mould. This method was also one of the earliest used in China to make paper.

The addition of a deckle box to the mould means that the newly formed sheets of paper can be removed before they are dry, so eliminating the need to have a mould for each sheet of paper made.

Deckle box and mould

Mould and Deckle, used for European/Western papermaking

The adapted mould is covered with a rigid, woven-wire mesh and is used with a wooden frame of the same size, but with high sides. This is called the deckle box. To use, they are clipped together, mould with mesh-side up and the deckle box placed on top of the mesh. While keeping them clipped together, deckle box on top and mould underneath, they are placed in a vat of water. The level of the water should be about halfway up the side of the deckle box. Enough pulp for one sheet of paper is poured into the deckle box and agitated by hand to help distribute the fibres evenly. Small frames work best with recycled wastepaper, half-stuff, linters, or wet lap, as the space is limited and very long fibres will get tangled around your hand. The frames are then lifted out of the vat of water and unclipped.

The freshly made sheet of paper is couched onto a felt or cloth and a stack of new papers separated by cloths is gradually built up; this is called a post of papers and it is now ready to be pressed. Once pressed, the layers are separated and the papers are hung up to dry while still attached to the cloths.

The only difference between the deckle box and mould, and the mould and deckle described below, is that the deckle box has high sides and is clipped to the mould when used, rather than held in place with your hands.

Mould and Deckle

This mould also consists of a wooden frame with a rigidly fixed, fine mesh, but the deckle is a wooden frame of the same size.

To use, the mould and deckle are held together, with the mould mesh-side up and the deckle placed on top. With a scooping action they are dipped into a vat containing water and pulp. The agitation of the pulp ensures that the fibres overlap in every direction, thus forming a strong bond, and the paper is lifted out in an even layer.

The mould and deckle are kept level while the excess water drains through the mould and the deckle is then removed to reveal the newly formed sheet.

This sheet is couched onto a felt or cloth and a stack of new papers separated by cloths is gradually built up. The papers are then ready to be pressed. Once pressed, the layers are separated and papers are hung up to dry while still attached to the cloths. This process can be used with fibres that are short or long, and the thickness of the papers made depends on how much pulp there is in the vat. The addition of a formation aid is recommended to help the even distribution of long plant fibres.

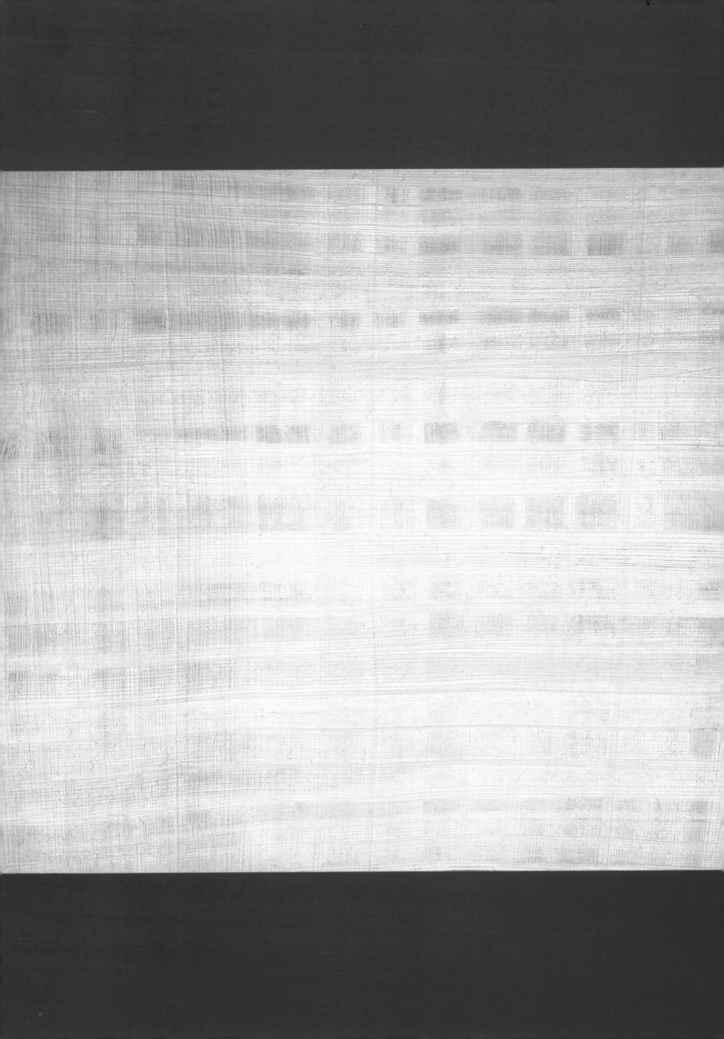

Is it paper?

Papyrus, amate and silkpaper may look like paper, but the different processes used to make them are not true papermaking methods, as previously described.

Papyrus

This is the first and best-known forerunner of paper. It was made in ancient Egypt, and the name 'paper' is derived from the plant used, *Cyperus papyrus*, a strikingly tall, aquatic plant, with stems reaching 3–8m (10–25ft) high. It grows profusely in the stagnant lakes and rivers in many parts of Africa, and the ancient Egyptians used it to create sheets for writing or drawing on. However, the process used to yield this 'paper' differed from that used for true papermaking: long strips of the layered stem and pith of papyrus were split and flattened, then laid side by side. Further layers were added, each one at right angles to the first, so they formed a cross-hatch pattern, and the strips were bound together with water. Once the desired thickness was reached, the layers were pressed and dried in the sun, then polished with a shell or other hard surface.

Left: **Papyrus**

Amate

Amate is bark 'paper'. It dates back to the Mayan and Aztec cultures but is still made today. The example of amate shown left is from Mexico and it is made in a similar way to papyrus. The rough outer bark is removed from strips of bark and these strips are then softened by boiling them in caustic soda. Once washed to remove the lye, the strips of softened inner bark are beaten flat to form 'paper'. To make larger pieces of amate, the strips are laid out with a slight overlap before being beaten flat.

Amate

Detail of amate

Silkpaper

Silkpaper

Silk fibre

The Dutch artist, Inge Evers, originally developed this technique in the late 1970s. Silkpaper is thin, like paper, with the distinctive uneven deckle-like edge found on handmade paper, but it is made with layers of silk fibre which have been glued together. Silk fibres are laid side by side on top of nylon net, then a second layer is added, but at right angles to the first, to make a cross-hatch pattern. Another piece of nylon net is placed on top of the silk fibre layers, glue is then pressed through the net, and the sandwich of silk and net is hung up to dry. When dry, the net is removed to reveal the silkpaper. The results are stunning, with vibrant shimmering silk fibre.

Some of the projects I show you will include books with silkpaper covers. Although it is a modern technique, the simple methods used are similar to those of the very first ancient methods of papermaking.

Basic materials and equipment for pulp papermaking

To make beautiful handmade papers you do not need expensive equipment. In fact the basic necessities are already present in most households.

Fibre for papermaking

Choose fibre that will make strong paper but will not damage your blender. As already mentioned in the chapter on Choosing Fibre (see page 7), I will be using bamboo half-stuff for all of the projects.

As a guide to how much fibre you will need, a finished A4 sheet of my handmade paper, made with bamboo half-stuff, has an average weight of 10g. If you are using wastepaper as your source of fibre, you should follow the basic papermaking techniques for sheet formation. All of the projects in this book can be made with wastepaper, but the finished results may not be as strong, so experiment.

Moulds and deckles

You will need a variety of sizes and shapes. These can be bought from a specialist shop, or you can make your own (see the 'How to Make a Mould and Deckle' on page 27).

You will need the following:

Materials and Equipment

Fibre for papermaking

Moulds and deckles

Vat (a plastic crate)

Blender

Kitchen cloths

Drip tray

Press and pressing boards

Bucket

Sieve

Bowls and measuring jug

Rubber gloves

Sponges

Hand towel

Size

Kettle

Glass or clear-plastic jar

Spoons

Turkey baster and empty squeezy bottle

String, paperclips and pegs

Vat

Use a plastic crate large enough to hold your mould and deckle and with room for your hands at each side. It must be at least as deep as the width (shortest side) of the mould and deckle to allow the pulp and water to move freely. One vat is sufficient, but you will need another one when you start to experiment with more decorative techniques.

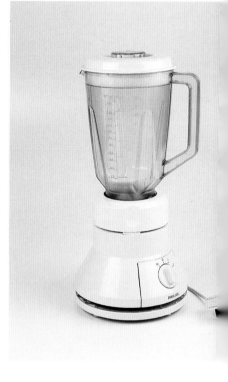

Blender

An ordinary kitchen blender to make your pulp is the small-scale alternative to the industrial-size beater. Do not use your blender for food preparation if you add dyes and pigments to your pulp.

Kitchen cloths

The 'all purpose' kitchen cloths shown right are an alternative to felt for couching your papers onto. They are thin, allowing your papers to dry quickly, very strong and durable, and can be washed, making them reusable many times. Alternatively, you could use cotton sheets cut down to size.

Drip tray

Use two lengths of wood to make a raised work surface inside a large shallow tray (right). This will stop water from going everywhere when couching or pressing your papers. The raised work surface and drip tray should be larger than the papers you plan to make.

Press and pressing boards

Your post of papers must be pressed to get rid of all the excess water before the paper is hung up to dry. If you don't have a standing press (as shown right), you can use heavy books and kitchen-scale weights. Alternatively, you could use your body weight. Whichever method you use, you should first place your post of papers between two large boards (called pressing boards) to even out the pressure. They can be made of any wood, as they are not in contact with water for long, or you could use Formica-covered boards. Either way, the boards must be left to dry out after each papermaking session.

Bucket

A bucket is used to pre-soak half-stuff and wastepaper in water, to soften it before processing with a blender.

Sieve

After you have finished making paper, pour away the contents of a vat through a sieve, to stop the leftover pulp blocking sinks. Also use a sieve for rinsing out excess dye from pulp.

Bowls and a measuring jug

Have bowls in a variety of sizes available for odd jobs, including pre-soaking small amounts of half-stuff in water before processing in the blender. Use the jug to transfer water from the vat to the blender, so you don't get the base of the blender wet.

Rubber gloves

To protect your hands, wear rubber gloves when using pulp that has been dyed or coloured with pigments.

Sponges

Use these to remove excess water when couching paper onto kitchen cloths. Also useful for mopping up any spills.

Hand towel

When using an electric blender, you must dry your hands first as a safety precaution.

Size

This is a type of glue which will make your paper less absorbent. Most suppliers of papermaking materials sell Aquapel but there are many alternatives, such as PVA, a white water-based glue; cellulose-based wallpaper paste; pure cellulose paste; spray-on clothes starch or cooking gelatine. I will be using cooking gelatine in all my pulp papers, with some additional size from spray-on clothes starch.

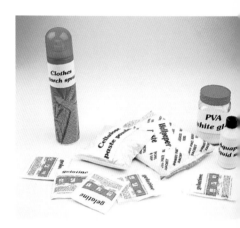

Kettle

For boiling water to dissolve gelatine size.

Glass/clear-plastic jar

Use a 500ml (1pt) jar to dissolve gelatine. One made of glass or clear plastic is best, so that you can see when the gelatine has dissolved completely.

Spoons

Have a variety to hand, for mixing gelatine size and for adding small quantities of pulp to the vat.

Turkey baster

Use a baster and empty squeezy bottle for decorative techniques with water, e.g. rain paper and water-cut paper.

String, paperclips or pegs

Make a temporary indoor washing line for hanging your papers up to dry. Hold the papers in place with pegs or paperclips.

How to make a mould and deckle

A mould and deckle set consists of two wooden frames of the same size. The mould has additional dowelling ribs for extra strength and also a mesh to form your paper on. Moulds and deckles can, of course, be purchased (see 'Useful Addresses' on page 181) but, if you prefer to make your own, this chapter tells you how to do it.

The mesh must be made of a material that does not rust; I used a fine aluminum mesh with 1 x 2mm ($\frac{1}{32}$ x $\frac{1}{16}$in) holes. You could use nylon net curtain as a substitute, but it must be stretched taut on the frame, and the texture should simulate a consistent, all-over fine mesh, and not have a distinctive pattern. The thickness of the wood used for making your frames should be taken into consideration when cutting it to the correct length, as it is the inside measurements of the mould and deckle frames that determine the size of the paper made.

Standard paper measurements

All of the papers made in the projects described use the 'A' series for measurements from the International Standards Organisation (ISO). However, handmade papers have a distinctively uneven, deckle edge, so they will vary slightly in size.

	Millimetres	Inches
A3	297 x 42	(11$\frac{11}{16}$ x 16$\frac{1}{2}$)
A4	210 x 297	(8$\frac{1}{4}$ x 11$\frac{11}{16}$)
A5	148 x 210	(5$\frac{7}{8}$ x 8$\frac{1}{4}$)
A6	105 x 148	(4$\frac{1}{8}$ x 5$\frac{7}{8}$)

I will be referring to all the papers in the projects by their 'A' series size.

The picture on the right clearly shows that if you fold a sheet of size A3 paper in half, you will get two sheets of size A4 paper. If you fold a sheet of A4 paper in half you will get two sheets of size A5 paper, while if you fold a sheet of A5 paper in half you will get two sheets of A6 paper.

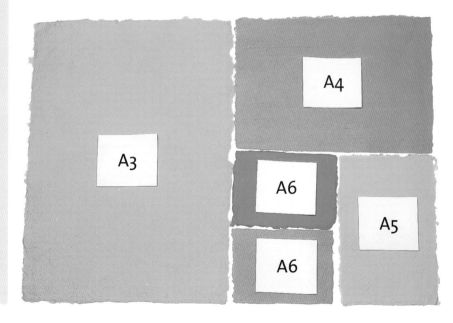

The table below shows you the dimensions of wood, dowelling and mesh, and the number of panel pins you will need for specific sizes of mould and deckle.

	Wood with 20 x 20mm (³/₄ x ³/₄in) in section with lap joints		Dowelling 8mm (⁵/₁₆in) diameter	Aluminium mesh or nylon net	Panel pins 20mm (³/₄in) long
A6 mould	2 off 145mm (5¹¹/₁₆in) long	2 off 188mm (7³/₈in) long	1 off 105mm (4¹/₈in)	170 x 210mm (6¹¹/₁₆ x 8¹/₄in)	4 off
A6 deckle	2 off 145mm (5¹¹/₁₆in) long	2 off 188mm (7³/₈in) long	No dowel required	No mesh required	4 off
A5 mould	2 off 188mm (7³/₈in) long	2 off 250mm (9¹³/₁₆in) long	2 off 148mm (4¹/₈in)	210 x 270mm (8¹/₄in x 10⁵/₈in)	4 off
A5 deckle	2 off 188mm (7³/₈in) long	2 off 250mm (9¹³/₁₆in) long	No dowel required	No mesh required	4 off
A4 mould	2 off 250mm (9¹³/₁₆in) long	2 off 337mm (13⁵/₁₆in) long	4 off 210mm (8¹/₄in)	270 x 360mm (10⁵/₈in x 14³/₁₆in)	4 off
A4 deckle	2 off 250mm (9¹³/₁₆in) long	2 off 337mm (13⁵/₁₆in) long	No dowel required	No mesh required	4 off
A3 mould	2 off 337mm (13⁵/₁₆in) long	2 off 460mm (18¹/₈in) long	5 off 297mm (11¹¹/₁₆in)	360 x 480mm (14³/₁₆in x 18⁵/₁₆in)	4 off
A3 deckle	2 off 337mm (13⁵/₁₆in) long	2 off 460mm (18¹/₈in) long	No dowel required	No mesh required	4 off

Method

The construction method used for all sizes of moulds and deckles is exactly the same so, once you have cut your pieces of wood to size and have all you require to hand, proceed as follows:

1 Apply waterproof wood glue on the lap joints to construct both the mould and deckle frames. Leave to dry.

For all sizes of moulds and deckles you need:

Waterproof wood glue

Yacht varnish

Paintbrush

Hammer

Scissors

Wall stapler and staples

Marker pen

Wood

Dowelling

Aluminum mesh or
 nylon net

Panel pins

(see table on facing page)

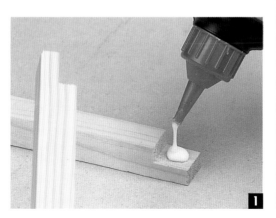

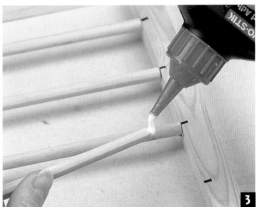

2 For extra strength, put
a panel pin in each joint.

3 Next, attach the
dowelling to form
strengthening ribs in the
mould. Position the pieces
so that they are level with
the top of the mould, and
equal distances apart. Fix in
place with waterproof wood
glue, then leave to dry.

4 Apply at least three
coats of waterproof yacht
varnish to all areas of the
mould and deckle, following
manufacturer's instructions
for both drying times
between coats, and brush-
cleaning methods.

5 Use the mould to hold
the aluminum mesh flat
while you cut it to size.

6 To make an exact fit, place the aluminum mesh on top of the mould and crease all the edges to form a guide for folding to size. Cut off excess bulk mesh at the corners.

7 Fold the mesh underneath, along the crease lines then, with ribs to the top, staple the mesh to the mould frame. If any of the staples are sticking up, use a hammer to push them in further.

8 The mould and deckle set is now ready to use.

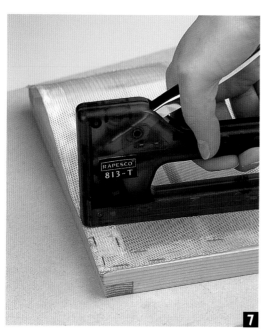

How to make an envelope-shaped deckle

Buy a commercial envelope that, when opened out flat, will fit on the deckle you want to use without touching the wooden frame. The most useful size of envelope to make is one that, when opened out flat, will fit onto a size A4 deckle.

To make an envelope-shaped deckle for use with an A4 mould, see 'Materials' panel, left.

Commercial envelope inside the deckle

Materials and Equipment

Commercial envelope, as described above

Foam board, 250 x 337 x 5mm (9⅞ x 13¼ x ³⁄₁₆in)

Marker pen

Scalpel

Cutting board

A4 deckle

Yacht varnish

Paintbrush

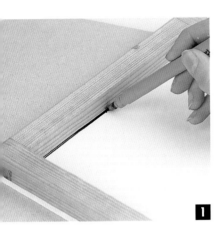

 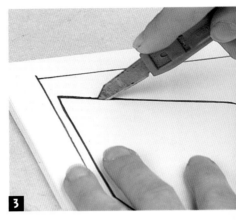

1 Use the A4 deckle to draw guidelines on the foam board.

2 Place the commercial envelope in the centre of the foam board, without it touching the guidelines, and draw around it.

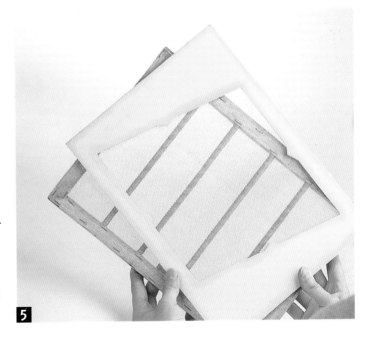

3 Use a scalpel to cut the envelope shape out of the foam board.

4 Apply at least three coats of waterproof yacht varnish to all areas of the envelope-shaped deckle. Follow manufacturer's instructions for both drying times between coats and for brush-cleaning methods.

5 Use the envelope-shaped deckle with an A4 mould.

Basic pulp papermaking techniques

Facing page: **Half-stuff**

This section guides you through the different stages in making paper by hand, using a mould and deckle. The same principles will apply, whatever the size or shape of paper you want to make. If you want to make envelope-shaped paper, use an envelope-shaped deckle (see page 32) instead of a wooden deckle frame.

When calculating the amount of half-stuff you need, remember that a finished sheet of my A4 handmade paper, made with bamboo half-stuff, has an average weight of 10g.

All of the projects will follow these basic techniques for making paper from pulp, but with additional decorative variations.

1

Preparing pulp

1 To take the strain off the blender, soak the half-stuff in a bucket of water for about 15 minutes, to help soften it.

Materials and Equipment

Mould and deckle, as required

Half-stuff, as required

Vat

Blender

Gelatine size (see page 38)

Kitchen cloths

Drip tray

Press

Pressing boards

Bucket

Sieve

Bowl

Measuring jug

Sponges

Hand towel

Kettle

Glass or clear-plastic jar

Spoons

String

Pegs

2 Half-fill the blender jug with cold water then add stamp-size pieces of pre-soaked half-stuff. Don't add too much, as this would strain the blender. As a guide, only blend one small handful at a time, as the pieces of half-stuff should be able to move around easily. Check that the half-stuff is not compacted in the bottom of the blender – if it is, the blades will be unable to move.

2

3 Make sure that the lid of the blender jug is firmly in place, then dry your hands before connecting it to the motor base (see 'Safety Note', below left).

4 Place one hand over the lid of the blender jug, then switch on, using the slower speed. If the blender starts to strain or move about, steady it with one hand and switch off. This could have been caused by a number of things:

a The blender jug wasn't connected to the motor base correctly. Disconnect and try again.

b There is too much half-stuff in the blender jug for the machine to process. Disconnect the blender jug from the motor base, remove the lid, take out some of the half-stuff and top up to halfway with cold water.

c The half-stuff has not been left soaking for long enough. Disconnect the blender jug from the motor base, remove the lid and pour the half-stuff and water back into the bucket to soak for longer.

4

Safety Note

When using an electric blender you should, for safety reasons, get into the habit of doing things in the same order: water, half-stuff, lid securely on, then dry your hands before connecting to the motor base and switching on the power. For most models there should be no need to have the blender jug on top of the motor base without fixing the lid securely in place first. Never put your hand into the blender jug while it is still fixed to the motor base. This operation should only be done by one person at a time, for safety reasons.

5 With the correct amount of half-stuff and water in the blender jug, switch the motor on at a slow speed for about ten seconds. The longer the motor is switched on, the smaller the half-stuff will become. You will learn to tell how long to run the blender motor, both by eye and by the feel of the pulp.

6 Disconnect the blender jug from the motor base and remove the lid. Once the half-stuff looks like a thick smooth porridge, it is called pulp.

7 Pour all the pulp into the vat and repeat the process, making a total of three blender jugs of pulp for an A4 piece of paper if using a vat large enough to take an A4 mould and deckle. For a smaller vat add less pulp and for a larger one add more.

8 Top up the vat with cold water. The level of water and pulp mix in the vat should be at least three quarters up the side of the mould.

Adding size to the vat

1 To make the paper less absorbent, and to give it added strength, a size is added to the vat of pulp. This is called internal sizing. If no size is added, then the papers made will be more absorbent and, if you draw on the paper with a brush and ink, it will bleed into the paper and look like the veins in a leaf. This is called 'waterleaf'.

The paper shown above left has been sized internally with gelatine, while the paper shown above right has not.

2 Pour 500ml (1 pint) of boiling water into a warmed glass, or clear-plastic jar.

Alternative Methods of Internal Sizing

Other substances can be used for internal sizing, such as one tablespoon of PVA white glue, one tablespoon of wallpaper paste, or one tablespoon of cellulose paste powder. These should all be dissolved in the same way as gelatine, i.e. before being added to the vat of pulp. Aquapel, a proprietory product, is already in liquid form, so one tablespoon can be added directly to the vat of pulp. To find the required absorbency, experiment with different quantities, to see which ones give additional strength to your paper.

3 Sprinkle in one tablespoon of gelatine, stirring all the time. Never add water to gelatine, as it will set quickly at the bottom of the jar.

4 Make sure that the gelatine is dissolved completely. If you can see small lumps, keep stirring until they are all gone. Any lumps of gelatine that get into the vat of pulp, however small, will act like a strong glue and may not be noticed until it is time to remove sheets of handmade paper that have been drying on a kitchen cloth, and they will not separate.

5 Pour the gelatine size into the vat of pulp.

Preparing a work surface

1 To prepare a raised work surface, place two lengths of wood inside a shallow drip tray, then lay one of your pressing boards on top of the wood lengths (see right).

2 Fold two kitchen cloths in quarters.

3 Dampen the cloths with water from the vat and place in the middle of the pressing board. This will make a padded surface to start a post or stack of paper on.

4 Spread a kitchen cloth out on top. This is the cloth your first sheet of paper will stay attached to until it is dry.

N.B. To have any effect, gelatine size needs about three weeks curing time once the paper is dry.

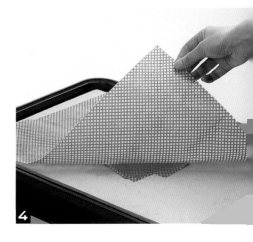

Pulling a sheet of paper

1 While you have been preparing your equipment, the pulp will have settled on the bottom of the vat. When you are ready to make (or 'pull') your first sheet of paper, stir the vat of pulp with your hand, making sure you reach the bottom.

2 Make sure the pulp mix is moving around when you form a sheet of paper. This makes for a stronger sheet, as the fibres will be overlapping each other in all directions, and than just side by side.

3 Hold the mould on the shortest sides, with the mesh uppermost, and place the deckle on top of the mesh. Keep the mould and deckle this way up while making paper, and hold them tightly together, so that pulp cannot escape between the frames. Do not touch the mesh, as this will result in holes in the paper where your fingers have been.

4 Submerge the mould and deckle in the vat of pulp, using a scooping action towards you.

To catch as much pulp as you can, and to make an even sheet of paper, only keep the mould and deckle in the vat of pulp for a matter of seconds.

5 Keeping the mould and deckle level, lift them out of the vat of pulp. Still holding the mould and deckle level, and directly above the vat, let excess water run off.

6 To help speed things along, take a sponge and, with one hand underneath the mould, soak up excess water, then squeeze the water from the sponge back into the vat.

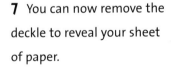

7 You can now remove the deckle to reveal your sheet of paper.

It is only with practice that you will be able to gauge how thick the sheet of paper you make will be.

If the finished sheet is too thin

a You may not have lifted all the pulp from the bottom of the vat when stirring it.

b When forming the sheet of paper, the mould and deckle may only have been submerged just under the surface in the vat of pulp, leaving most of the pulp behind in the vat.

c The mould and deckle could have been held together too loosely, so that the pulp escaped between the gaps.

d You can't make paper from this vat of pulp forever. The more sheets of paper you make, the more pulp you will be removing from the vat each time. Add more pulp, but only one blender-full at a time, to help keep the thickness of the sheets even.

If the finished sheet is too thick

The pulp will slide off the mould when you try and couch it onto the cloth. This means that there is too much pulp in the vat. Remove excess pulp by dipping a sieve in the vat.

All paper may look uneven when still on the mould, but pressing the post of papers will remove any lumps. If the mould has been lifted out of the vat at an angle, then the pulp will be thick at one end and the mesh of the mould visible at the other. To start again, remove the pulp from the mould by dipping it, mesh-side down, into the vat. Do not scrape it off with your hand, as the pulp will form into little lumps. Agitate the pulp in the vat by stirring it with your hand before making the next sheet of paper, and make sure that the mould and deckle are kept level when lifted out of the vat.

Using the European-style mould and deckle, it is the amount of pulp in the vat that determines the overall thickness of the sheets of paper made: e.g. one initial blender jug of pulp in a vat that is large enough to make size A4 sheets of paper will result in very thin sheets of paper. Four or more blender jugs-full of pulp in the same size vat will result in very thick and lumpy sheets of paper.

Couching a post of paper

1 To couch your paper onto the kitchen cloth, rest the mould on one edge and, with the same action as a door hinge, ease the frame down with the other hand.

2 Keep the mould steady with one hand and, using a firm pressing action with a sponge in your other hand, ease the paper off the mould and onto the cloth. Squeeze the excess water from the sponge back into the vat.

If the couching is unsuccessful

Unsuccessful couching may occur because the paper has torn, or is very thin. Both problems can be solved by replacing the mould on the cloth in exactly the same place, then filling your sponge with water and flooding the mould. This will help to loosen the pulp from the frame and tears will rejoin. Press firmly as before, but do not get rid of the excess water before removing the mould.

  If the couching still doesn't work, then remove the pulp from the mould by dipping it, mesh-side down, into the vat. Do not scrape the pulp off with your hand, as that would cause it to form into little lumps. Add more pulp before making another sheet of paper, then refer to the section on 'When to add more pulp to the vat' (overleaf) and repeat the procedure.

3 Again using the door-hinge action, hold down one side of the mould as you lift up the opposite side. Be brave: doing this slowly doesn't help. Keep the mould registered in the same position until you are sure your sheet of paper is satisfactory. If not, replace the mould in exactly the same position and refer to 'If the couching is unsuccessful' panel, left.

4 Once you have couched successfully, cover the newly formed sheet of paper with another cloth and couch your next sheet of paper onto this cloth.

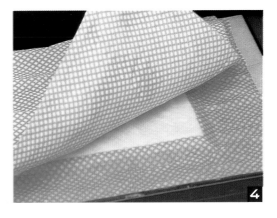

When to add more pulp to the vat

The more sheets of paper you make, the more pulp you will use from the vat. As a guide, you may be able to make five sheets before your paper starts to look very thin and you can see the mesh through it while it is on the mould. It is then time to add more pulp.

To keep your papers at an even thickness:

a Only add one blender jug-full of pulp at a time.

b Use water from the vat for processing new half-stuff in the blender. Don't take water from the tap, as this would dilute the size too much and your papers will be waterleaf (see 'Adding size to the vat', page 38).

When to add more size to the vat

Each time you make a sheet of paper, you remove water as well as pulp from the vat so, when making a lot of sheets of paper, the level of water in your vat will drop.

If the level of water has dropped by half, making it difficult to pull a sheet of paper successfully without the mould and deckle bumping on the bottom of the vat, you should top up with fresh water from the tap. By doing this you will dilute the size solution so much that it will not have any useful effect on your paper, so you should now add more size to the vat, i.e. another tablespoon of gelatine dissolved in 500ml (1 pint) of boiling water.

Pressing the post of papers

Pressing the post of papers not only gets rid of excess water, but it also helps the fibres to form a stronger bond, thus making the paper stronger.

Obviously, the more sheets of paper you make, the higher your post of papers and cloths will be. As a guide, start by making no more than ten sheets of paper in your post.

Try and keep the sizes of paper in each post the same, to avoid unwanted marks when pressed. You should also remove the initial folded kitchen cloths before pressing.

Cover the final sheet of paper in the post with a cloth, to protect it and prevent it from sticking to the pressing board. A dry cloth is easier to spread than a wet one, even though it will absorb water as soon as it touches the newly formed sheet of paper. Cover with the other pressing board.

Use one of the following three methods for pressing your post of paper:

a Place the papers, still between the pressing boards in the standing press, and slowly apply pressure by tightening. The more papers you have in the post, the longer it will take to press them. I press large quantities of paper overnight, slowly applying more pressure. There will be a lot of water expelled from the paper, so make sure that the standing press is inside a drip tray but on top of wooden strips to raise it up out of the collecting water.

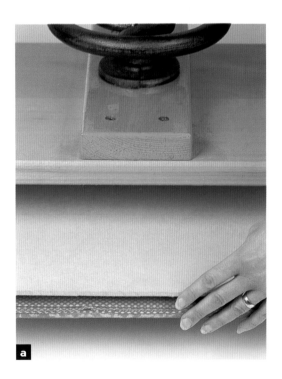

b If you do not have a standing press, use heavy kitchen weights and large books to apply pressure and expel water from the post of paper. Again there will be a lot of water, so keep the post of papers in the drip tray, but on top of wooden strips, to keep it out of the collecting water. If you can, press for at least two hours, but the longer the better.

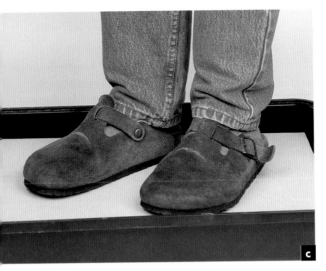

c Lift the post of papers onto the floor, still between the pressing boards and on wooden strips in the drip tray. Apply pressure by standing on the pressing board. This works best with about ten sheets of paper in your post, so you don't have to stand there all day. If you do not have a drip tray, keep the post of papers between the pressing boards, take them outside and stand on them. If you are indoors, then use newspapers or cotton sheet to soak up the water.

Drying your paper

1 Lift off the top pressing board and the first protective cloth.

2 Each cloth will now have a sheet of damp paper attached to it. Carefully take a corner of the top cloth and peel it back slowly.

3 Remove the cloth, with paper still attached (see picture, right), and hang on an indoor washing line. Continue until all the cloths with papers attached are hung up, then dry slowly, overnight (see 'Tip', on facing page).

4 To separate the dry paper from the cloth, work on a flat dry surface for support, with the paper visible. Lift one corner of the paper then slide your hand flat between the paper and the cloth, to loosen it.

5 To flatten the paper, press it overnight with weights between dry pressing boards.

Your paper is then ready to use.

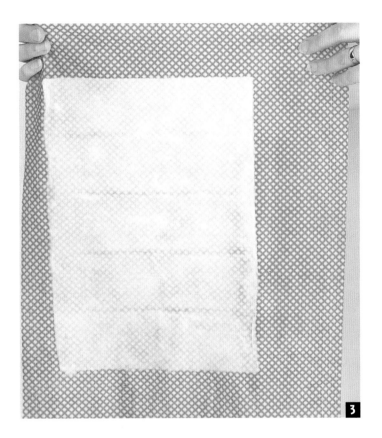

3

Surface texture of paper

Your paper will have the same texture as the surface it has been dried on, in this case, kitchen cloths. If you want very smooth paper, couch it directly onto Formica or perspex instead of cloth, and leave there until dry. Alternatively, you can simply iron your papers flat, once dry.

External sizing

If you have made paper without internally sizing it, or you want to add extra size to sheets of paper that are already dry to make them less absorbent, they can be sized externally with a spray-on starch for clothes. Iron the starched paper to fix, repeating the process until you get the desired effect. This will also make your papers stiff.

Tip

Make sure that your indoor washing line is not positioned above any electrical items or above anything that you don't want to get wet. Although the papers have been pressed, they may still drip. Don't hang them up near any strong direct heat as the papers will curl up when drying. If you hang the papers up outside, any slight breeze will cause them to fall off the cloth.

Colouring pulp

Colouring pulp is easy, can add an exciting new dimension to your stationery and, provided you follow the guidelines below, it can safely be carried out in the home. Have fun, as all three different methods of adding colour will give you brilliant results.

Lightfastness test

It can be interesting to do a small experiment on your coloured papers. To do this, take two sheets of identical paper. Place one of them in a sealed envelope and tape the other one to a south-facing window. Compare the sheets of paper at weekly intervals for a month or so, to see if the colour has faded. If there is no colour difference, then the paper is said to be lightfast. You can also do the same with a fluorescent light source.

pH test

The pH value is a scale of 1 to 14 on which the acidity or alkalinity of a solution is measured – pH7 being neutral, below this, acid, and above, alkaline. If you want your papers to last, a pH reading between 6 and 8 is recommended, as both strong alkaline and acid solutions could have an adverse effect on your paper over time.

When using pigments, dyes and other additives, always ask the suppliers for advice before purchasing, or use pH test strips to do your own tests.

Health and Safety Warning

Take sensible precautions when using dyes or pigments, especially if working in the kitchen:

- Cover the work area with a plastic sheet.
- Wear rubber gloves and a protective apron at all times when using pigments and dye.
- To avoid inhalation of pigments and dyes that are in a dry powder form, always wear a dust mask. These substances can be detrimental to your health and some are carcinogenic.

- When using pulp that has been coloured with dyes or pigments in a vat for sheet formation, wear gloves to prevent skin irritation.
- Make sure there is no food left uncovered, and work on the draining board of your sink, so that any spills can be cleared up.
- Do not use any equipment that is used in the preparation of food for colouring with dyes or pigments; keep equipment specifically for use when dyeing, and clean it thoroughly after use.
- For further advice, please refer to individual manufacturers' health and safety precautions.

Making brightly coloured papers

White half-stuff does not make exciting paper on its own but you can make brightly coloured papers by colouring the pulp before forming it into sheets. The following three methods will all give you brilliant results.

Materials and Equipment

Brightly coloured wastepaper

Half-stuff

Bucket

Blender

Pre-Coloured Wastepaper

1 Soak the half-stuff and coloured wastepaper in a bucket to soften. Half-fill the blender jug with water and add stamp-size pieces of pre-soaked half-stuff and coloured wastepaper. The paper should be able to move freely in the blender jug.

2 Follow safety guidelines for operating the blender, i.e. put the lid on, dry your hands, then fix the blender jug to the motor base and switch on for about ten seconds, or until the pulp is an even colour. The coloured pulp is now ready for use.

3 You can get a variety of shades by varying the ratio of white half-stuff to coloured wastepaper. The more colours of wastepaper, the more fun you can have.

Note

The proportion of coloured wastepaper you use depends on the colour you want to achieve. Also, bear in mind that, if you use more than half wastepaper, the resulting handmade paper will be considerably weaker than if only a quarter of wastepaper is used with half-stuff.

Because this technique uses wastepaper, it is best to process just enough pulp for immediate use. If you dry the coloured pulp, it will have to be processed through the blender again when you want to use it and, each time the pulp is processed in the blender, the fibres will get shorter and the resulting papers weaker.

The lightfastness of the colour in your paper will depend on the type of colourant used in the original wastepaper. If this is unknown, you can do a lightfastness test and keep records for future reference (see page 49).

Pigments

Pigments are coloured rocks (earth tones) or synthetic particles (all colours) that have been ground up into a fine dust. You can buy them in powder or liquid form. A process similar to that of a magnet is used to attach pigment to pulp. Cellulose plant fibre, in this case half-stuff, has a negative or anionic charge, which is increased when processed in a blender. Pigments on their own have no charge, so a retention agent with a positive or cationic charge is added to the pigment. A case of opposites attract. Pigments are very lightfast.

Note

Retention agent is added to the fibre to give it a positive charge. It is available from suppliers of papermaking materials/equipment (see pages 181–2).

Materials and Equipment

Pigments

Retention agent

Half-stuff

Weighing scales

Measuring jug

Measuring syringe

Rubber gloves

Bucket

Blender

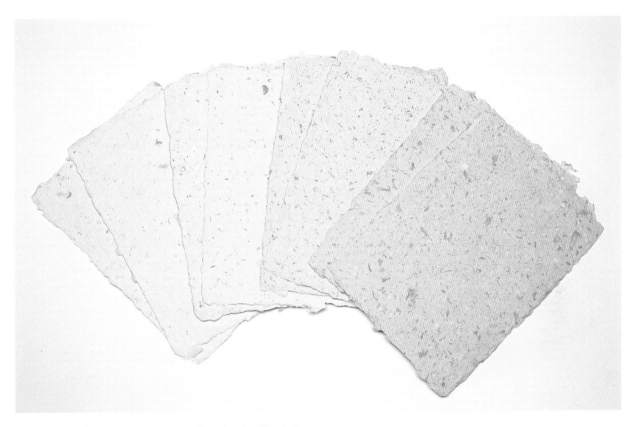

Paper coloured with pigments at 1%, 2%, 5% and 10% solutions

If you want to repeat a colour at a later date, it is important to keep paper samples along with a record of method and quantities of pigment used. The table below will show you how to use the retention agent at strengths with neutral pH values, so they don't need to be rinsed out of the pulp at the end of the dyeing process. Pigments have a neutral pH value and can be used at any strength without the need to be rinsed out of the pulp.

	Pigment (liquid form)	Dry weight of half-stuff	Retention agent
1% pigment	1ml (¼ tsp)	100g (3½oz)	1g (½ tsp)
2% pigment	2ml (½ tsp)	100g (3½oz)	1g (½ tsp)
5% pigment	5ml (1 tsp)	100g (3½oz)	1g (½ tsp)
10% pigment	10ml (2 tsp)	100g (3½oz)	1g (½ tsp)

To use pigment at 1%:

Wear an apron and rubber gloves at all times. If using pigments in dry powder form, also wear a dust mask.

1 Measure 1ml (¼ tsp) of liquid pigment and dilute with a little water in a jug.

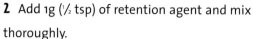

2 Add 1g (½ tsp) of retention agent and mix thoroughly.

3 Weigh out 100g (3½oz) of dry half-stuff and it soak in a bucket, to soften. Half-fill the blender jug with water and add stamp-size pieces of pre-soaked half-stuff. The paper should have room to move freely in the blender jug.

Follow safety guidelines for operating the blender: put the lid on, dry your hands, then fix the blender jug to the motor base and switch on for about ten seconds, or until the pulp is smooth. Pour the pulp into a bowl. Repeat until all 100g (3½oz) of the half-stuff has been processed.

4 Next, stir in the diluted pigment and retention agent.

5 The coloured pulp is now ready to use, as there is no need to rinse.

Again, I recommend only processing enough pulp for immediate use, as, if you dry the coloured pulp, it will have to be processed through the blender again when you want to use it. Each time it is processed, the fibres will get shorter and the resulting papers weaker.

You can mix different colours of pigmented pulp together like paints to make a more personal range of coloured papers. The more colours of pigments you have, the greater the range of papers you will be able to make.

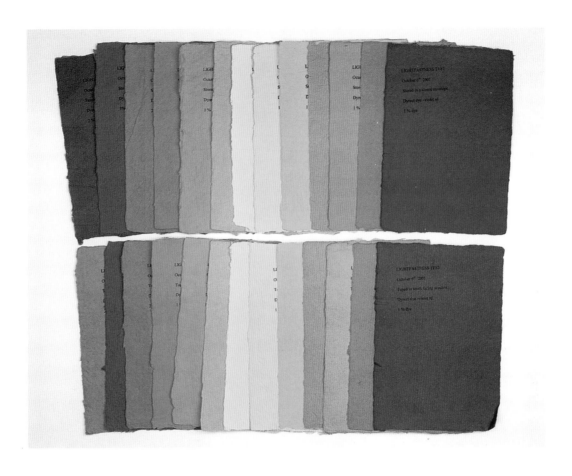

Direct Dye

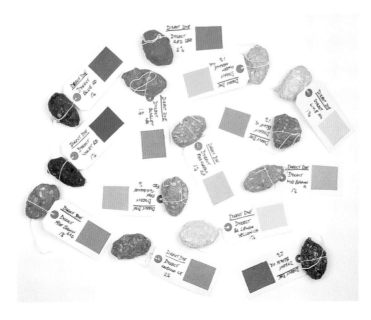

Direct dyes are chemical colourings that have a direct affinity to the cellulose in natural fibre. The quality of lightfastness is very good, but varies from colour to colour (direct dye will run out if washed, but that doesn't apply here).

The top row of handmade papers shown above has been kept in a sealed envelope away from any light. The bottom row shows identical handmade papers, but they have been taped to a south-facing window for one month and some of the direct-dye colours have faded slightly.

If you want to repeat a colour at a later date, it is important to keep paper samples along with a record of method and quantities of direct dye used. The table below shows you how to use the direct dye and rinse aid at strengths with neutral pH values, so that they do not need to be rinsed out of the pulp at the end of the dyeing process.

However, the quantities of salt used will be alkaline and must be rinsed out of the pulp. This is a slow process, as the pulp will retain water. To eliminate the need for rinsing, you can omit the salt, as it is only there to improve the colour take-up of dye in the pulp. This will vary from colour to colour, so experiment.

Materials and Equipment

Items that will come into contact with direct dye should be made from stainless steel, plastic, enamel, wood or glass, as these will not cause a chemical reaction which would affect the colour.

Direct dye

Common salt

Rinse aid with a neutral pH value

Half-stuff

Dust mask

Rubber gloves

Weighing scales

Measuring jug

Measuring syringe

Kettle

Bucket

Sieve

Spoons

	Direct dye	Dry weight of half-stuff	Common salt	Rinse aid
1% direct dye	1g (½ tsp)	100g (3½oz)	20g (6 tsp)	5ml (1 tsp)
2% direct dye	2g (1 tsp)	100g (3½oz)	20g (6 tsp)	5ml (1 tsp)

To use direct dye at 1%:

Wear an apron and rubber gloves at all times, and wear a dusk mask while the direct dye is in dry powder form.

1 To weigh out 1g of direct dye, place it on the scales in its plastic jar, take a reading and then, with a spoon, remove the dye until the scale reads 1g less than before. Alternatively, measure ½ tsp of the dye.

2 In a measuring jug, add a little boiling water to the measured direct dye and mix it to a thick paste, removing any lumps.

3 Add 20g (6 tsp) of salt to the paste and top up with enough boiling water to dissolve it.

4 Half-fill a bucket with boiling water and add the dye mix. This is called the dye bath.

5 Soak 100g (3½oz) of half-stuff until soft, then tear it up into hand-size pieces. Add it to the dye bath, using a wooden spoon to ensure that the half-stuff is completely covered by water and can move around freely. Add more boiling water if necessary, then leave until cold.

6 Add 5ml (1 tsp) of rinse aid to the dye bath and mix thoroughly. This should make the dye bath become clearer and will also help to reduce the amount of rinsing at the end of the dye process. Leave for 20 minutes. Patches of white undyed half-stuff will still be visible at the end of the process, so don't panic.

7 Use the sieve to rinse the half-stuff until the water is almost clear.

8 Squeeze out excess water, then dry the dyed half-stuff for later use. This may take a week or so.

100g (31/2oz) of dry dyed half-stuff

9 Before using the dry-dyed balls of pulp, soak them in water to soften, then process them in a blender. Any white half-stuff still visible in the dyed ball of pulp will disperse to form an even colour once blended.

You can mix different colours of dyed pulp together like paints to make a more personal range of coloured papers. The more colours of direct dye you have, the greater the range of papers you will be able to make.

All of the half-stuff used in the projects in this book has been coloured with direct dye.

Blue/white direct-dyed handmade papers

Blue/lime direct-dyed handmade papers

5 Add a second layer in the same way, but with the silk fibres running at right angles to the first. This cross-hatching will make your silkpaper strong.

6 Once two layers of silk fibres are complete, cover them with the second piece of net.

7 Add 5g of cellulose paste powder (or wallpaper paste) to 500ml (1 pint) of water in a measuring jug. Stir vigorously as you pour, to keep it smooth. Leave for 30 minutes or so, until the paste thickens and any lumps have dissolved.

8 The consistency should be similar to that of double cream. If it is too thin, gradually add more paste powder. If too thick, dilute with water.

9 Hold the net steady and pour on some of the paste.

10 With a sponge, press the paste through the net. Add more paste as necessary.

11 When you have sponged all over the net, scrape off any excess paste and replace it in the jug. Holding two corners, turn the layers of net and silk over.

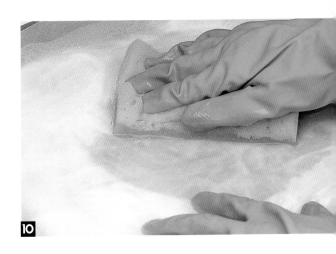

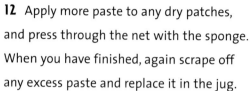

12 Apply more paste to any dry patches, and press through the net with the sponge. When you have finished, again scrape off any excess paste and replace it in the jug.

13 Take two corners and hang it on a temporary indoor washing line to dry overnight. Position the plastic sheet on the floor directly below the line to catch the drips of adhesive.

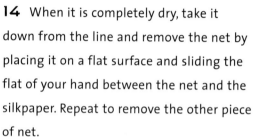

14 When it is completely dry, take it down from the line and remove the net by placing it on a flat surface and sliding the flat of your hand between the net and the silkpaper. Repeat to remove the other piece of net.

15 Your silkpaper is now ready to use.

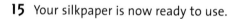

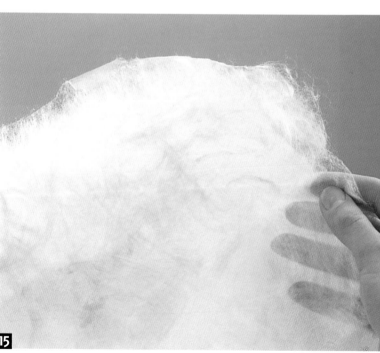

Adding Strength to Silkpaper

1 To add strength to the silkpaper, so that it can be used for book covers, apply iron-on vilene before you cut the silkpaper to the sizes required. The vilene will be on the inside of the cover and only visible when the book is opened.

If you want to keep the feathery edge of the silkpaper, the vilene should be just smaller than the silkpaper.

If you want the silkpaper to have a straight edge, the vilene should cover the silkpaper completely. Iron on the vilene.

2 Measure out the size of paper required by drawing on the vilene, then cut to size.

Materials and Equipment

Silkpaper

Iron-on vilene

Iron

Pen

Scissors

Strengthened silkpaper

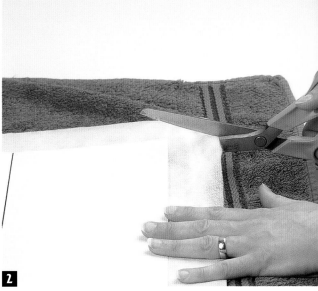

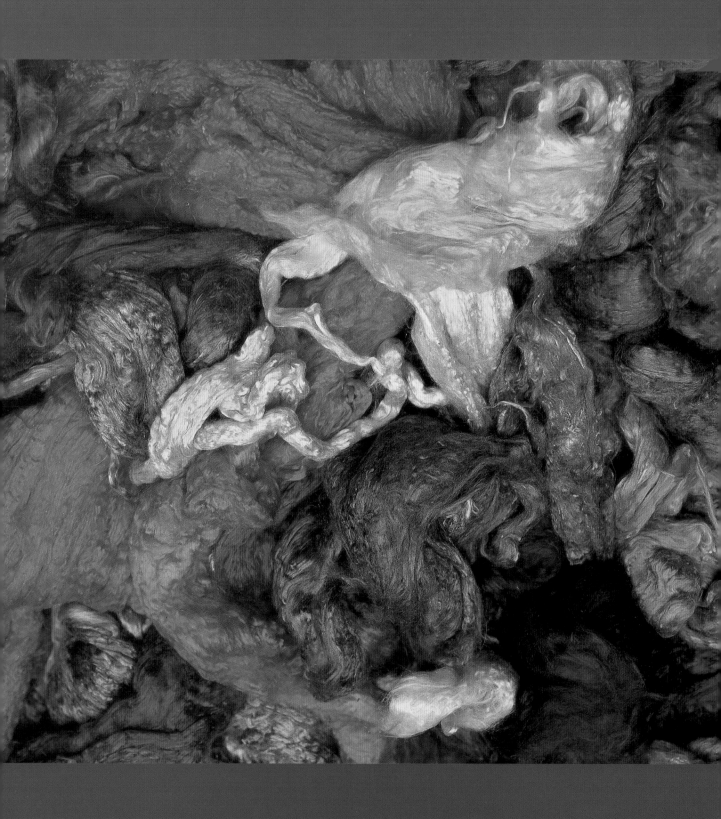

Colouring silk fibre with direct dye

Materials and Equipment

Items that will come into direct contact with direct dye should be made from stainless steel, plastic, enamel, wood or glass, as these will not cause a chemical reaction affecting the dye colour.

Direct dye

Household white vinegar

Silk fibre

Dust mask

Polythene food wrap

Liquid soap

Kettle

Hotplate

Steamer, or large pan with lid and close-fitting metal sieve

Heavy-duty plastic protective sheet

Tongs

Shallow plastic tray. Mine is 38 x 26 x 2cm (15 x 10 x ¾in)

Spoons

Bowl

Rubber gloves

Weighing scales

Measuring jug

Measuring syringe

Colouring silk fibre is not complicated, and you don't have to be a chemist to achieve fantastic results. Take a look at the list of equipment you need and you will see that most, if not all, will already be in your kitchen. If you follow the guidelines below, the process can be carried out safely in the home.

When using direct dye to colour silk – which is an animal fibre – white vinegar is added instead of common salt, to make the solution acidic and to improve colour take-up. If you want to repeat a colour at a later date, it is important to keep silk samples along with a record of method and quantities of direct dye used. The table overleaf will show you how to use the direct dye at 1% and 2% strengths.

	Direct dye	Dry weight of silk fibre	White vinegar
1% direct dye	1g (½ tsp)	100g (3½oz)	30ml (6 tsp)
2% direct dye	2g (1 tsp)	100g (3½oz)	30ml (6 tsp)

To use direct dye at 1%:

1 Weigh out 100g (3½oz) of dry silk fibre, and soak it in a bowl of water containing a few drops of liquid soap and 10ml (2 tsp) of white vinegar. The soap-and-vinegar solution acts as a wetting agent, which will help dye to penetrate the silk fibre more easily. Agitate the silk fibres to make sure that they are thoroughly wetted.

2 The shiny silk fibre will go dull once it has been properly wetted, a process that may take about ten minutes.

3 Squeeze out excess liquid.

4 To weigh out 1g of direct dye, place the small plastic jar containing the dye on the scales, take a reading and then, with a teaspoon, remove the dye until the scales read 1g less than before.

In a measuring jug, mix the measured direct dye with a little boiling water to make a thick paste, stirring to ensure that there are no lumps. Add 20ml (4 tsp) of white vinegar, then top up to 500ml (1pt) with cold water.

To make an exciting range of intermediate tones, mix several individual colours as described above, each one in a different container.

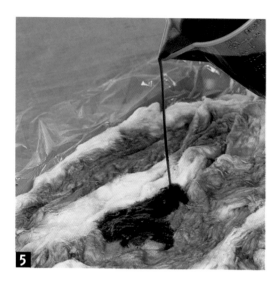

5 Protect your work surface with a plastic sheet, then place a tray on top of the plastic. Roll out enough polythene food wrap to completely cover the tray with generous overlaps and make a sealed surface. Lay out the silk fibre. Pour some of the direct dye solution onto the wetted silk fibre, without causing a flood (see picture, top right). If using a second colour, dye the remainder of the silk fibre with that. This simple technique is called rainbow dyeing. You need just enough liquid to saturate the silk fibre and no more, as it should soak in completely, with no excess.

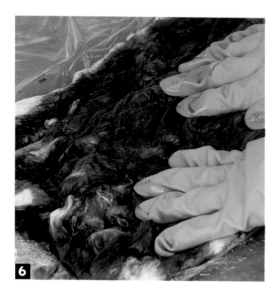

6 Press firmly to ensure that the dye has penetrated the silk fibre (picture, above right).

7 Frequently turn the silk fibre and tease it open to make sure that the dye has soaked in evenly.

8 Fold in all the sides of polythene food wrap Roll up to make a completely sealed package (see picture top left, overleaf).

9 Next, the dye must be fixed with heat. Half-fill a pan with water and rest the sieve inside. The water should not touch the sieve. Place the sealed package of polythene food wrap, dye and silk fibre into the sieve.

10 Cover with a close-fitting lid. Bring to boiling point, then simmer for at least 30 minutes to steam-fix the dye. Don't let the pan boil dry.

11 The liquid inside the polythene package should now be almost clear, although there may be some colour-staining in the boiling water from dye dripping out of the polythene food wrap.

12 Use tongs to lift the dyed silk out of the sieve and then rinse the package in a bowl of cold water.

Tip

Leftover dye solutions can be stored in clearly labelled opaque plastic containers for later use. The dye will separate out and settle at the bottom of the container, so shake well before using, then add a little boiling water to dissolve any dye granules that have formed.

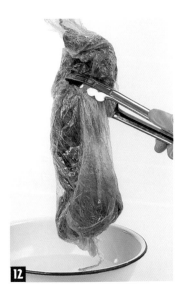

13 Remove the polythene food wrap and change the water in the bowl once, to give the silk fibre a good rinse. Squeeze out the excess water then hang up the dyed silk fibre to dry completely before using to make paper. As the silk fibres are very strong, it can be draped over an indoor washing line to dry.

14 The dyed silk fibre is now ready to use for making paper.

Dyed silk fibre

Silkpaper made from lime green and turquoise silk fibre

Tip

To use direct dye at 2%, follow the directions in the table for the quantities of direct dye, dry weight of silk fibre and vinegar. You will see that the only weight that has changed is the amount of direct dye, which has doubled. The directions for use are exactly the same as for direct dye at 1%.

Decorative envelopes

How to form an envelope

Envelopes, writing paper and greetings cards are things we use almost every day. To make yours much more personal, for sending to special friends, simply follow these projects.

1 Place the card in the centre of your handmade envelope, lining up with the notches formed in your handmade paper. Using a bone folder, fold in the sides to form the envelope shape.

2 Keep the card in place while you apply glue to two of the edges. Fold up to fix the sides in place, press firmly, then remove the card.

3 Your envelope is now ready to use.

Use paper glue, sealing wax, double-sided tape, or coloured, gummed paper shapes to seal your envelope once your letter is inside.

Materials and Equipment

Handmade envelope shape (see page 32–3)

Paper glue

Bone folder

Card: the size of your envelope when formed

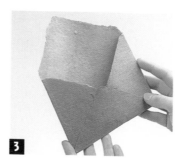

Seaweed inclusions

Project:

Envelopes with seaweed inclusions

Inspiration:

Seaweed

Add an aura of the seaside to your stationery by adding seaweed to the standard vat of paper pulp. Collect the seaweed at low tide in rock pools, as it will be much cleaner than the seaweed found on the highest tidelines, which are not washed by the sea very often.

Look for a variety of colours and shapes of seaweed but avoid any that are very chunky, or look unhealthy. I've used maiden's hair, *Corallina officinalis*, dulse, green hairweed, grass kelp and sea lettuce. Seaweed comes to life when it is free to float about in water, and it is best to use fresh seaweed to replicate this effect in paper. You don't need to collect

Materials and Equipment

Standard pulp papermaking equipment, plus

Bucket

Water hose

A4 mould

A4 envelope-shaped deckle

White pulp

A selection of fresh, cleaned, seaweed

Coloured gum-paper shapes

a vast amount – a couple of handfuls from each variety selected will do.

Once you get the seaweed home, give it a good rinse in tap water, to get rid of sand and shells, which would damage the blades of your blender.

If left to dry naturally, seaweed will shrivel up and become an uninteresting hard lump so, if you want to save the seaweed for use later, it will need to be pressed. To do this, put it between two sheets of blotting paper, with a pressing board on either side, then place it under pressure, using kitchen weights or books if you don't have a press. It will take a couple of weeks to press flat, so keep checking it and change the blotting paper as required if it is very damp, to stop the seaweed from going mouldy. Like pressed flowers, seaweed will lose its colour with time but you will still be left with interesting shapes and textures in your paper.

1 As mentioned above, collect your fresh seaweed in rock pools, where it will be cleaned regularly by the tide.

2 Rinse the seaweed with tap water, being sure to remove any objects such as shells, sand, or fishing line, that may damage your blender.

3 Separate the seaweed into groups of similar colour or shape, which will give you a greater variety of finished papers.

Note

Anything you add to the vat of paper pulp will end up in your blender when you prepare more pulp. If you are using dulse, do cut it into small pieces with scissors first, otherwise it will get tangled around your blender blades. Remove any stray pieces of seaweed that may get tangled around the blade in the blender jug before you connect it to the motor base and switch on. Couch, press and dry, following basic papermaking techniques (see page 35).

Tip

Adding variety

Each sheet of paper need not be the same. When you are ready to make your next sheet of paper, add one of the other types of seaweed to your vat of paper pulp, to create variety.

4 Take a small handful of seaweed. I used the bright green varieties: grass kelp, sea lettuce and green hairweed. Carefully separate the seaweed out and pull it into shorter strands, so that it floats freely in the vat of paper pulp (see picture, facing page, bottom right).

5 Gently remove the long strands of seaweed and extra paper pulp that may have strayed outside the new sheet of paper, before you remove the envelope-shaped deckle from the mould. Take care not to tear your newly formed envelope.

6 To complete, refer to 'How to form an envelope', stages 1–3, on page 77.

7 Seal the envelope, once the letter is inside, with coloured gummed paper shapes, sealing wax, double-sided tape or paper glue.

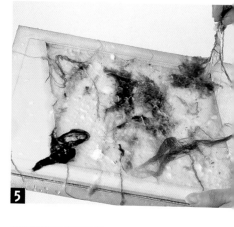

Seaweed papers and envelopes hanging up to dry

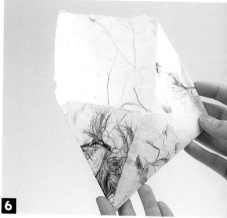

Seaweed paper envelope

Wastepaper inclusions

Project:

Envelopes with photocopied fish design inclusions

Inspiration:

Box of Fish

Fish come in all shapes and sizes with some quite fantastic patterns. You don't have to catch your own fish to explore them further, just look at fishing magazines and books for inspiration. Keep your designs bold and simple and, with the aid of a photocopier, you can soon produce multiple images. Unlike computer-printed designs, photocopied images won't run when submerged in water. Choose pastel shades of paper to show your designs at their best.

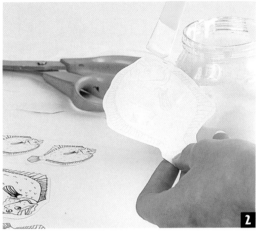

Materials and Equipment

Basic pulp-papermaking equipment, plus

Photocopied images

A4 mould

A4 envelope-shaped deckle

White pulp

A4 pale blue photocopy paper

A4 white paper

Pen

Scissors

Paper glue

1 Use bold, strong lines to draw a simple fish design on size A4 paper. With the aid of a photocopier, make many copies, each time reducing your fish in size.

2 Cut out and arrange the shoal of fish on a sheet of A4 paper. Glue them in position.

3 Photocopy them onto both sides of an A4 sheet of pale blue paper.

4 Soak the sheet in a bucket of water, then process in a blender, leaving large pieces of paper still visible. Add to a vat of white pulp. Alternatively, if you want very large pieces of fish designs in your pulp, simply tear up and add directly to the vat of pulp.

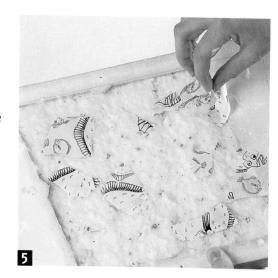

5 Pull a sheet of envelope-shaped paper. Carefully remove any pieces of paper that are on top of the envelope deckle.

6 Couch, press and dry, the sheet, following basic papermaking techniques (see page 35).

The final envelope

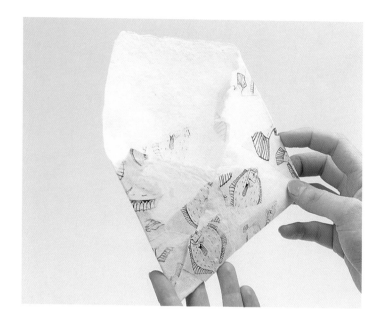

Rain paper

Project:

Envelopes with rain paper patterns

Inspiration:

Fishing boats

If you look at boats with peeling paint, you will see some great and unexpected colour combinations. As the paint flakes off, many layers of different colours are revealed: some are clashing, others may be different shades of the same colour. I have used large droplets of water to recreate a heavy downpour of rain, making large holes in the paper as they hit the surface. Large holes will allow you to see through the layers of paper, like a boat in much need of a new coat of paint. If you want smaller holes, use droplets of water flicked with your fingers and the effect will be much more subtle.

 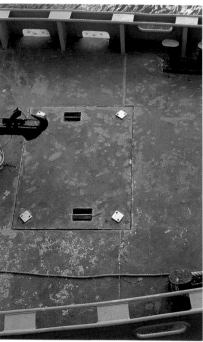 

1 Add turquoise pulp to the vat – half the amount of the you would normally use – and make a very thin sheet of envelope-shaped paper. Sponge off excess water, but do not remove the envelope deckle.

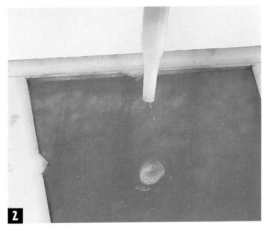

2 Fill the turkey baster with water from the vat then, holding the mould and envelope deckle above the vat, squirt strong jets of water through the paper to make the 'rain'. This will disperse the pulp, making holes in it. The bigger and more holes, the better the effect. If you prefer small holes, and a more subtle effect, just flick water on with your fingers. The thinner the paper is, the easier it will be to make your patterns.

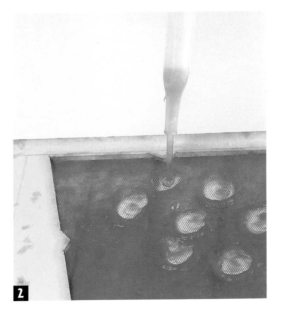

Materials and Equipment

Basic pulp papermaking
  equipment, plus

A4 mould

A4 envelope-shaped deckle

Turkey baster

Turquoise pulp

Dark blue, handmade
  envelope

3 Sponge off excess water and remove the envelope deckle.

4 Take the dry, dark blue, handmade envelope and place it on top of the turquoise rain paper. Use a dry envelope, because it is easier to position than a damp one.

5 Couch onto kitchen cloths (see picture, top right). The two sheets of paper will stick together without glue.

   Press and dry, following the basic instructions on pages 44–6.

The finished rain paper envelope

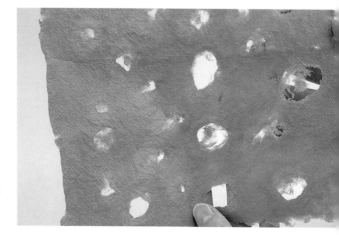

You can make sheets of rain paper without using an extra supporting sheet of paper. However, although they are very decorative, they are not practical

Printmaking

Project:

Envelopes with block-printed designs

Inspiration:

Anchors

Anchors have very simple, but easily recognizable, shapes. This quality makes them ideal for reducing to postage-stamp size, without loss of detail, which is perfect for a personal seal on your letters and envelopes. Alternatively, you could add your own patterns to a larger anchor design and use it as a greetings card or as an image to be framed.

1 Draw your anchor design on the polystyrene printing tile with a ballpoint pen. Place it on a wooden board and cut out the anchor with a scalpel. To avoid injury, do not cut towards your body and keep all fingers out of the way.

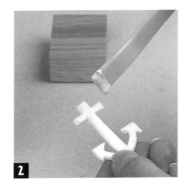

2 To make the stamp, glue the anchor onto a small block of wood.

3 Roll out the bright red printing ink on a tray.

4 Roll in both directions to spread the ink out evenly on the roller.

Materials and Equipment

Yellow handmade paper envelope

Red water-based block printing ink

Ink roller, or paintbrush (see 'Note' on facing page)

Ink tray (if using roller)

Polystyrene printing tile

Ballpoint pen

Scalpel

Wooden cutting board

Small wooden block

Glue

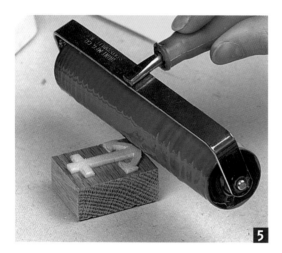

5 Pressing firmly, roll ink onto the anchor, without touching the wooden block.

6 The anchor should be evenly coated.

7 Position the stamp carefully on the envelope.

8 Press down firmly, holding the envelope still.

9 While still holding the envelope steady, lift up the stamp to reveal the printed anchor. Leave the envelope to dry overnight.

10 Once dry, your envelope is ready to use.

Note

If you do not have a roller, you can apply the ink to the block design with a flat-bristled paintbrush, no more than 1.5cm (⅝in) wide. Apply a small amount of printing ink to the block in as few strokes as possible. Avoid repeatedly going over areas already inked-up, as you will remove existing ink rather than apply more. Aim for a smooth finish on the surface of the block. The larger the design, then the larger the paintbrush you can use. Water-based inks dry quite quickly, so move swiftly.

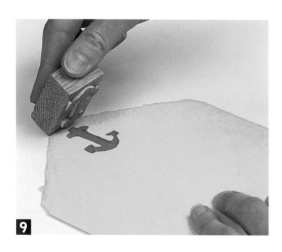

Other uses for polystyrene printing tiles

1 Polystyrene printing tiles can also be used for relief printing, by pressing in a design with a ballpoint pen. It can be fun experimenting with simple mark-making and patterns.

2 Use the roller to cover your patterns with an even coat of printing ink.

3 Lay your paper on top of the inked area and press firmly all over with the flat of your hand.

4 Lift up your paper to reveal the patterns, then leave the paper to dry overnight.
    You can use your stamp and relief print many times if you carefully wipe off any excess printing ink. Always wash the roller and printing tray in water once you have finished.

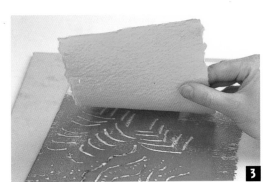

Stationery gallery

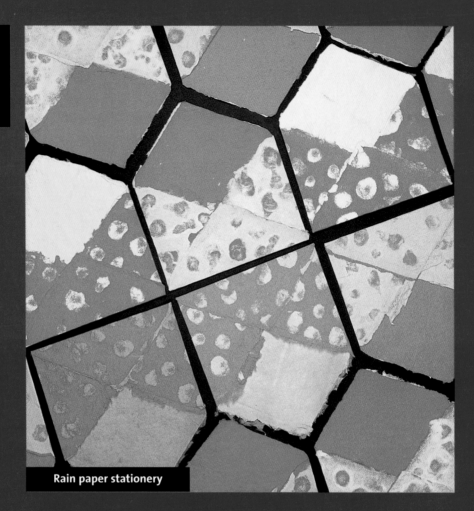

Rain paper stationery

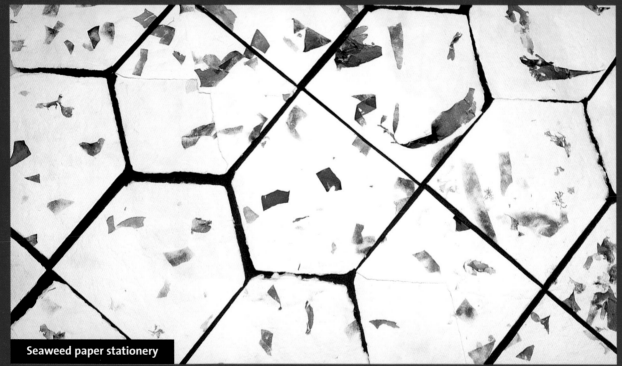

Seaweed paper stationery

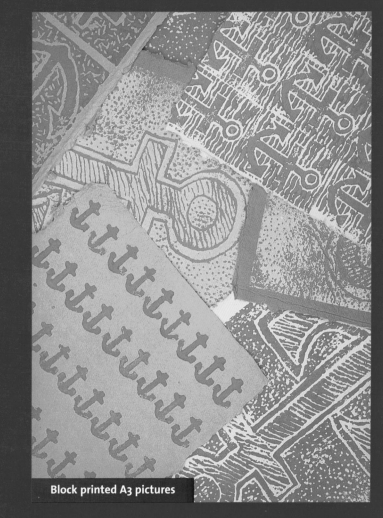

Block printed A3 pictures

Seaweed paper stationery

Fish designs

Photocopy fish stationery

Rain paper stationery

Block printed stationery

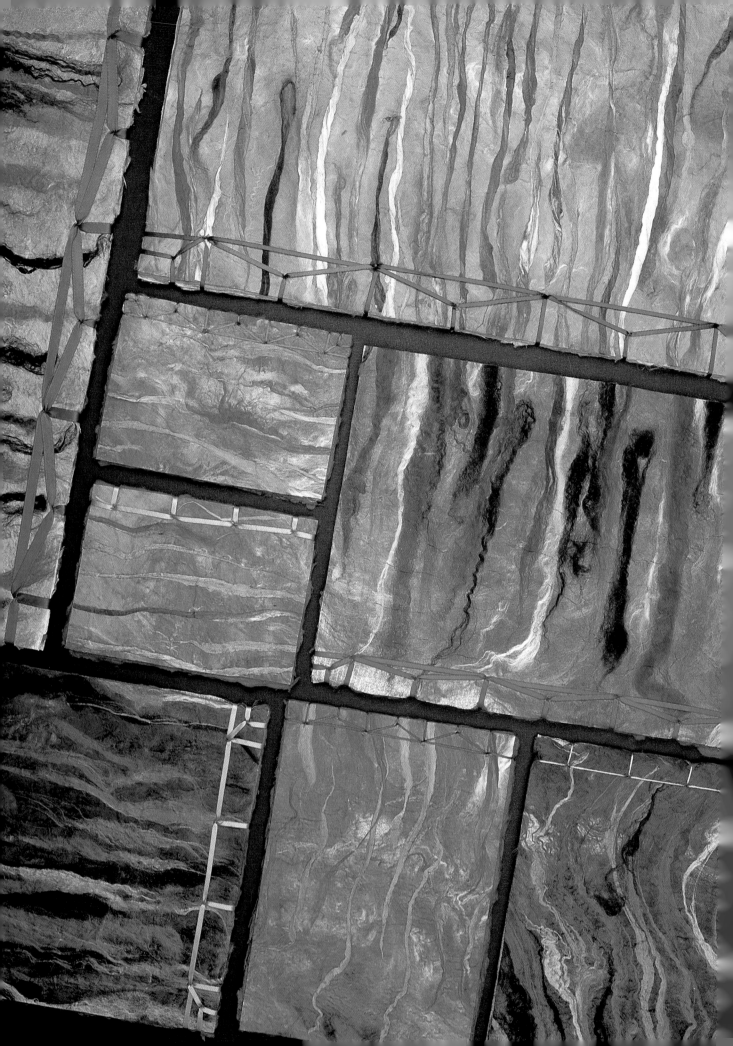

Japanese Stab Binding: Four-hole Stitch

Japanese stab-stitching is, perhaps, one of the most decorative bindings you can find, but, follow the instructions below, and you'll be amazed at how easy it is to make your own books. Even if you don't use your own handmade paper for the pages, the decorative stitch on the binding will add that extra touch of class.

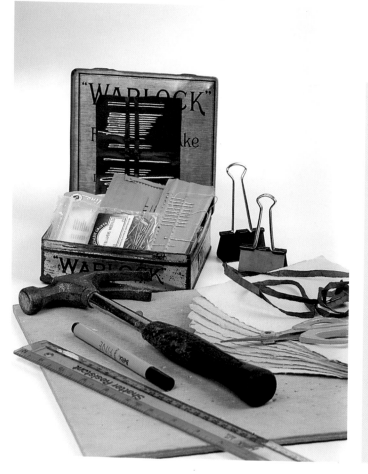

Materials and Equipment

Cover: (lime green) 2 sheets of A5 handmade paper

Pages: (turquoise) 8 sheets of A5 handmade paper

Paper string (see step 4)

Sewing needle

2 bulldog clips

1 large nail

Hammer

Wooden board

Scissors

Pen

Ruler

Method

1 Clip all your pages together in the correct order for binding. Whichever edge you choose to bind is called the spine. I am using the width of the book (shortest side) for the spine here.

Mark out the stitch holes 1.5cm (⅝in) in from the spine. As handmade paper has a distinctive uneven deckle edge, use your eye when lining up the ruler and marking out the stitch holes.

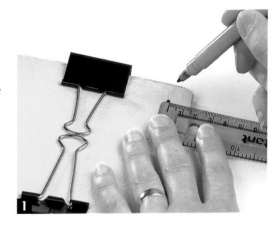

2 With your ruler parallel to the spine and 1.5cm (⅝in) in from the spine edge, mark out four stitch holes: 1.5cm (⅝in), 5.5cm (2⅛in), 9.5cm (3¾in), 13.5cm (5⅜in).

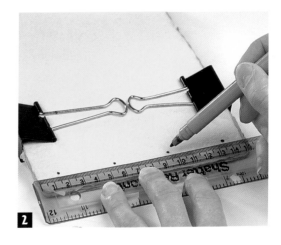

3 Strike a hole at each of these points with the hammer and nail.

4 Your paper string should be six times the length of the spine. For this book, that means six times 15cm (5⅞in), i.e. 90cm (35½in).

5 Position the book with the spine facing you, and keep it the same way up for all the stitching. Split open the pages at the spine and start sewing at the second hole from the right, pushing the needle from the opening through to the top. Leave a short tail of paper string.

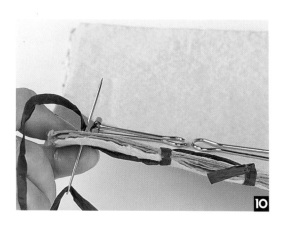

6 At the same hole, take the needle from the bottom to the top, making a stitch over the spine. Pull the paper string tight.

7 Push the needle down through the next hole to the left.

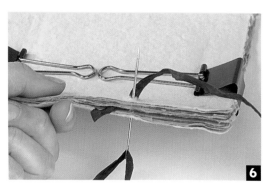

8 At the same hole, take the needle from the top to the bottom, making a stitch over the spine. Pull the paper string tight.

9 Push the needle up through the next hole to the left.

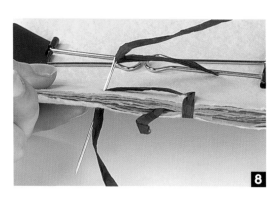

10 At the same hole, take the needle from the bottom to the top, making a stitch over the spine. Pull the paper string tight.

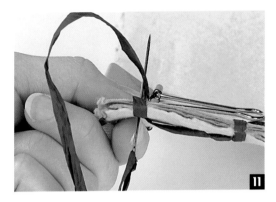

11 At the same hole, take the needle from the bottom to the top, making a stitch around the side of the book. Pull the paper string tight.

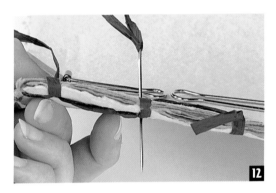

12 Now you will be working back, from left to right. Push the needle down through the next hole to the right.

13 Push the needle up through the next hole to the right. This will be the same hole you started from.

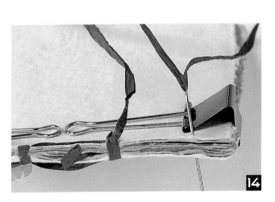

14 Push the needle down through the next hole to the right.

15 At the same hole, take the needle from the top to the bottom, making a stitch over the spine. Pull the paper string tight.

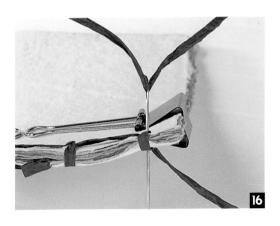

16 At the same hole, take the needle from the top to the bottom, making a stitch around the side of the book. Pull the paper string tight.

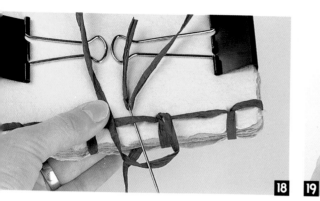

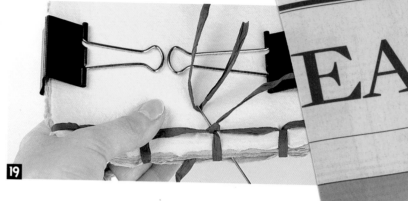

17 Push the needle up through the next hole to the left. This will be the same hole that you started from.

18 Make a loop with the paper string. Thread the needle under the first stitch to the left and through the loop of paper string. Pull the paper string tight.

19 Push the needle down through the same hole and bring it out, splitting the pages in the spine.

20 Trim the ends.

21 With the point of your scissors, push the ends back in through the split pages in the spine.

22 Your Japanese stab book is now complete.

Silkpaper

Project: A5 Japanese stab book with a silkpaper cover

Inspiration: Mackerel

The grey-blues and purples of mackerel shimmer and sparkle in the sun, making a rich combination of colours, and these bold stripes of colour make the mackerel stand out from the rest of the catch. The luscious sheen of silk, when dyed these colours, is the obvious choice of fibre to recreate the character of this jewelled fish.

Materials and Equipment

Cover:

Basic silkpaper making equipment, plus

Blue, grey, black and purple silk fibre

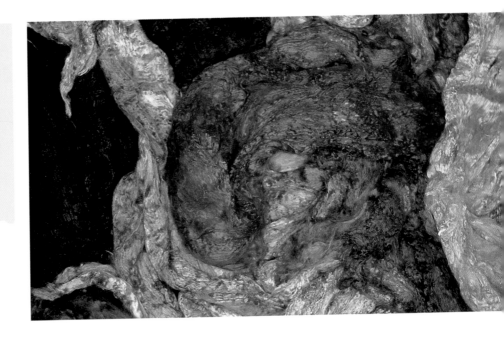

To make
the cover:

1 Follow the basic silkpaper making technique to lay out two layers of blue silk fibre (see page 61).

2 Place long slithers of different-coloured silk fibres on top.

3 Follow the basic silkpaper making techniques to glue, dry and remove from the net (see page 64).

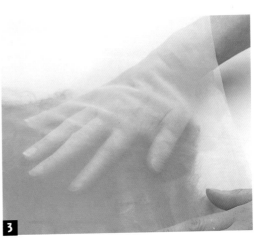

Materials and Equipment

Binding:

Basic Japanese stab binding equipment, plus

Cover: two sheets of A5 silk paper, strengthened with vilene

Pages: eight sheets of purple A5 handmade paper

Binding: blue silk embroidery thread

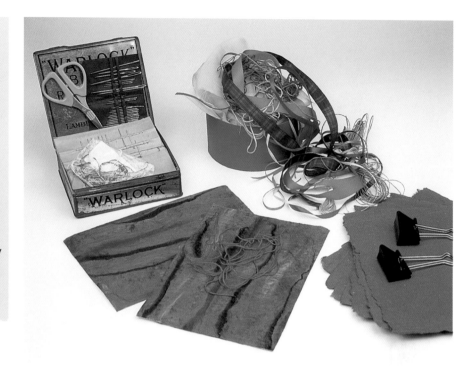

Method

Follow the Japanese stab method of binding (see page 99).

Reinforce with vilene then cut into two pieces, each A5 size

A4 Silkpaper

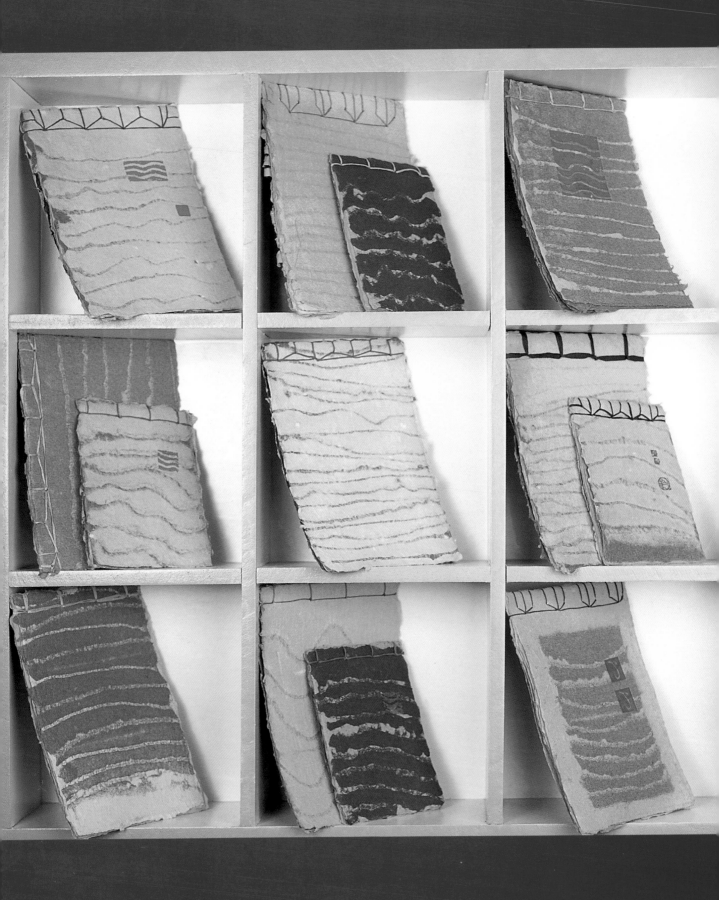

Watercut designs

Project:

An A5 Japanese stab book with watercut design cover

Inspiration:

Tidelines

Drawing with water is not the most obvious medium or technique to choose, but it is the best one to create soft lines when perforating a newly formed sheet of paper.

Ripples in shallow waters have a very similar look when their elusive lines wash up on the sandy shore, breaking up as soon as they have been created.

For this project, I chose yellow ochre and blue papers, to give the illusion of moving water on a sandy beach.

Materials and Equipment

Cover:

Basic pulp-papermaking equipment, plus

A5 mould and deckle

Squeezy bottle

Yellow ochre pulp

Two sheets of dark blue A5 handmade paper

To make the cover

1 Add to the vat half the amount of yellow ochre pulp you would normally use, and make a very thin sheet of paper.

2 Sponge off excess water, then remove the deckle.

3 Fill the squeezy bottle with water from the vat. Hold the paper, still attached to the mould, above the vat and draw lines with strong jets of water through the paper. The thinner the paper is, the easier it will be to make your patterns.

4 Sponge off the excess water again.

5 Take a dry sheet of ready-made dark blue paper and place it on top of the yellow ochre watercut paper (a dry sheet of paper is used because it is easier to position).

Materials and Equipment

Binding:

Basic Japanese stab binding equipment, plus

Cover: two A5 sheets of blue and yellow ochre watercut paper

Pages: eight sheets of A5 handmade dark blue paper

Binding: dark blue paper string

6 Couch onto kitchen cloth. The two sheets of paper will stick together without glue.

7 Press and dry, following the basic papermaking techniques (see instructions on pages 43–6).

Method

For this book I have used the length (longest side) as the spine. With your ruler parallel to the spine and 1.5cm (⅝in) in from the spine edge, mark out four stitch holes: 1.5cm (⅝in), 7.5cm (2⅞in), 13.5cm (5 in), 19.5cm (7⅝in).

Follow the Japanese stab method of binding (see pages 99–103).

Watermarks

Project: An A5 Japanese stab book with watermarked pages

Inspiration: Feathers

A watermark is created when the copper wire design displaces pulp, making the resulting paper thinner. Watermarks can only be seen when the papers are held up to the light. If your paper is very thick, or the wire used is too thin, the watermarks will not be visible.

When you look at feathers closely, you will see that they have very different profiles: some are thin and elegant, others are short and wide. If you line them up side by side to compare, you may be surprised at the variety of forms.

The outline of a feather can be drawn with very few lines, which makes designing a watermark very easy. The fewer the sections, the fewer the joints that will have to be soldered together. Most of my feather watermarks have been made from no more than five different pieces of copper wire.

Materials and Equipment

Basic pulp-papermaking equipment, plus

A5 mould and deckle

Pen

Sketchbook

Pliers

Copper wire

Solder

Solder machine

Protective, plain wooden board (see 'Note' below)

Dark blue pulp

1 To ensure that your watermark fits onto the A5 mould, use the inner size of the A5 deckle to mark out a guideline on your sketchbook, then draw a simple feather design in this shape.

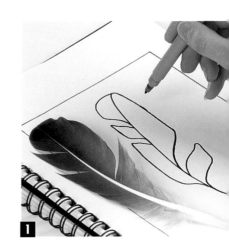

Note

The purpose of the wooden board is to protect your worktable when using a soldering iron, so any flat piece, such as plywood, will do, as long as it is large enough to take all of the soldering equipment and to work on it at the same time. However, it should not be varnished, as this would give off fumes if burnt, nor should it be covered in plastic or Formica.

To make the watermark

2 Cut and bend the copper wire so that it matches your design exactly.

3 Solder the pieces together.

4 Stitch the feather watermark onto the mould with strong cotton thread. Tie any knots on the back of the mould.

5 Your mould with a feather watermark is now ready to use.

6 The raised wire displaces pulp on the mould.

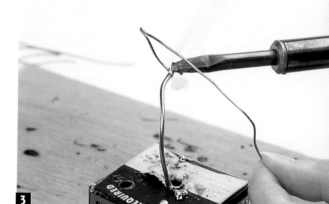

Copper feather watermarks, ready for stitching onto the mould

Materials and Equipment

Binding

Basic Japanese stab binding equipment, plus

Cover and pages: ten sheets of dark blue, A5 feather-watermarked paper

Binding: tartan ribbon

Method

For this book the length (longest side) has been used as the spine.

With your ruler parallel to the spine and 1.5cm (⅝in) in from the spine edge, mark out four stitch holes: 1.5cm (⅝in) 7.5cm (2⅞in) 13.5cm (5¼in) 19.5cm (7⅝in)

Your tartan ribbon should be six times the length of the spine. For this book, that means six times 21cm (8¼in), i.e. 126cm (49½in).

Follow the Japanese stab method of binding (see pages 99–103).

The feather watermark is only visible when held up to the light

Landscape paper

Project:

An A5 Japanese stab book with a landscape design cover

Inspiration:

Cliffs

High cliffs, formed from layers of pink and yellow sandstone, rise from the depths and stand proud, while weather-beaten grass and heather cling to the edge. The contrast of smooth stone and rough grass provides the inspiration for these layers.

Collect flat, dried grasses and leaves, or empty the contents of your old spice and herb jars to add texture to the paper and build your own landscapes.

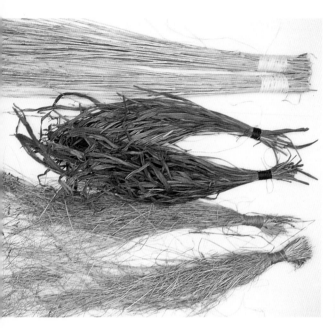

To make the cover

Materials and Equipment

Cover

Basic pulp-papermaking
 equipment, plus

A5 mould and deckle

3 vats

White pulp

Yellow ochre pulp

Rust-coloured pulp

Dried grasses

Scissors

1 Add short lengths of dried grasses to a vat of white pulp and make an A5 sheet of paper. Do not use long lengths of grass, as they might become tangled around the blades of the blender when you add more pulp. I used montbretia from my garden (see right).

2 In the second vat, make half a sheet of yellow ochre-coloured paper, by only dipping the mould and deckle in the vat halfway. Couch directly on top of the white grass paper to build up layers of colour and texture (see picture, facing page, top left).

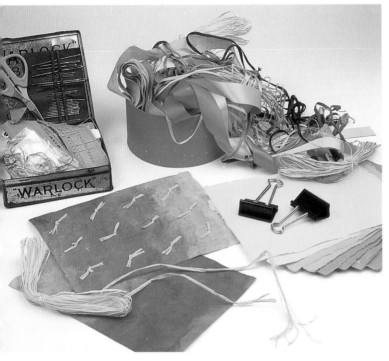

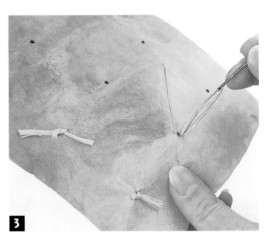

Materials and Equipment

Binding:

Basic Japanese stab binding
  equipment, plus

Cover: two sheets of A5 red,
  orange and yellow silkpaper,
  one with decorative
  stitching, strengthened
  with vilene

Pages: eight sheets of
  red/orange/yellow A5
  handmade pulp paper

Binding: yellow raffia

3 Mark where knots are
going to be, spacing them
equal distances apart,
remembering to leave a
space for the binding
(see picture middle right).
Make a single stitch with
yellow raffia at each mark,
then tie a simple knot.

Method

Follow the Japanese stab
method of binding (see
pages 99–103).

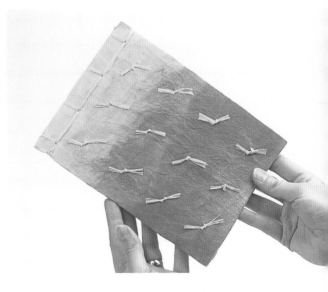

Japanese stab-binding gallery

Silkpaper books: layers of colour

Landscape paper b

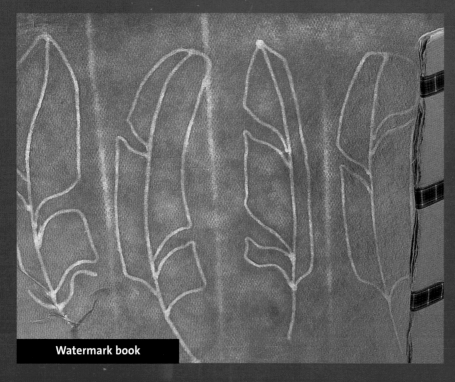

Watermark book

Watercut books

Watermark books (watermark visible when held up to light)

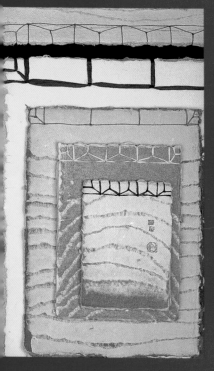

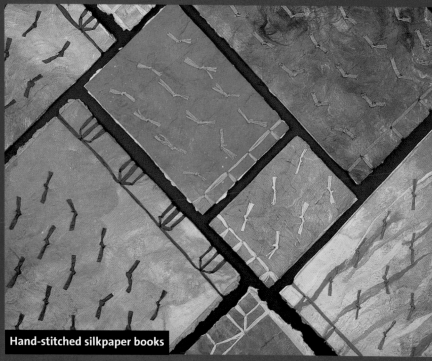

Hand-stitched silkpaper books

Single-section Binding

This is a very simple technique and there is very little chance of you going astray, as only three holes are necessary for this method of single-section binding. This simplicity allows you to concentrate on choosing decorative ribbons, to make your books much more flamboyant.

Materials and Equipment

Cover: one A4 sheet of lime-green, handmade paper

Pages: three A4 sheets of turquoise handmade paper

Ribbon

Sewing needle

Two bulldog clips

One large nail

Hammer

Wooden board

Scissors

Pen

Ruler

Card for hole template

Bone folder

1 Use the bone folder to firmly fold the pages in half (this folded edge is the spine of your book and the edge that will be stitched). Next, place all the pages in a pile, one inside the other, with the cover page on the outside.

2 Cut a length of ribbon three times the length of the spine. For this book, that means three times 21cm (8¼in), i.e. 63cm (24¾in) long, which will give you the option to tie off with a bow at the end of stitching.

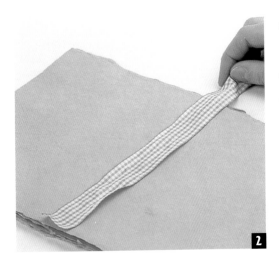

3 Clip the pages together and open out to see the centre page. Make a card template to show the length of the spine divided into four equal sections, then mark a line to represent the spine. Strike a hole at each of the three crosses with the hammer and nail. Remove the card template before binding.

4 Thread the needle with the ribbon, then start by pushing the needle in through the central hole, from the outside of the book to the inside. Leave a tail of about 10cm (4in).

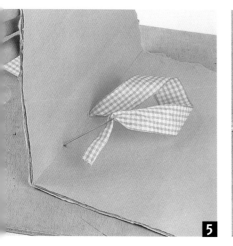

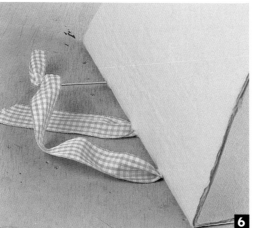

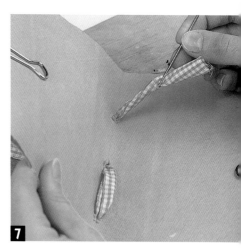

5 Push the needle out through the end hole nearest to you.

6 Push the needle back in through the last remaining hole.

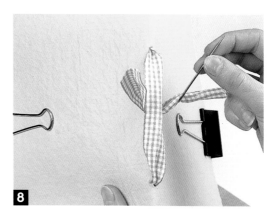

7 Push the needle back through the central hole to the outside of the book.

8 The needle should appear at the opposite side to the length of ribbon running along the length of the spine.

9 Tie off with a simple knot or bow and trim the ends of the ribbon, and your three-holed single-section book is then complete.

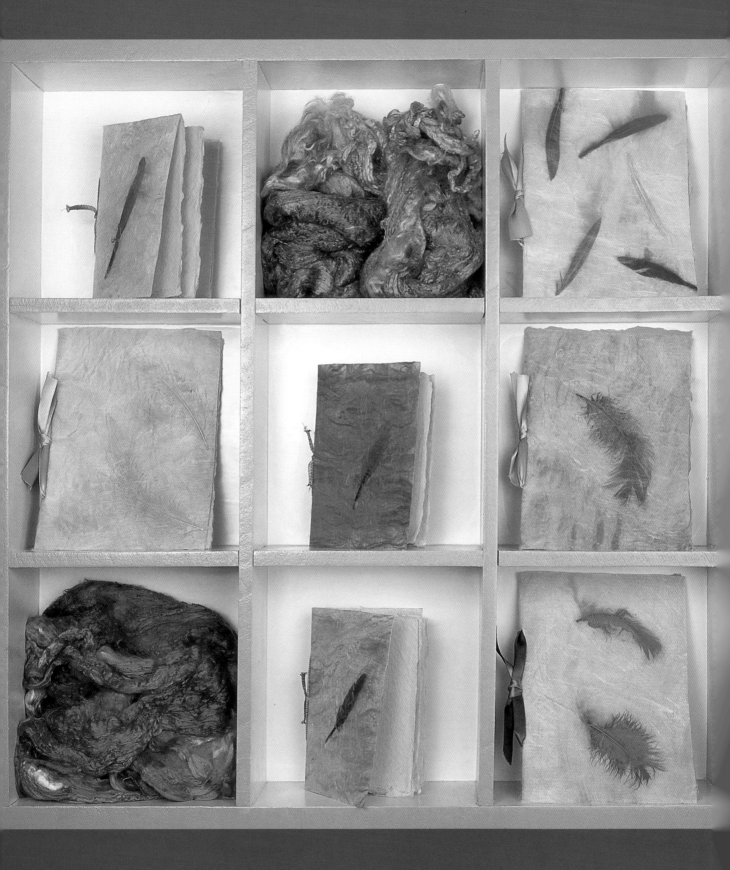

Silkpaper with feather inlay

Project:

An A5 single-section book with a silkpaper cover and trapped feathers

Inspiration:

Sea-fishing feathers, also known as mackerel feathers

The fishing feathers below have been dyed bright colours to attract fish in the dark deep waters off the coast. Whether you dye your own feathers or buy them dyed, they will add an exotic touch to your silkpapers. Some colours may bleed out of the feathers when glued in the silk, giving some surprising results. Remove any hooks before adding feathers to the silk fibre, so you are not the one to get hooked!

Materials and Equipment

Cover:

Basic silkpaper making equipment, plus

Two lime-green feathers

Turquoise silk fibre

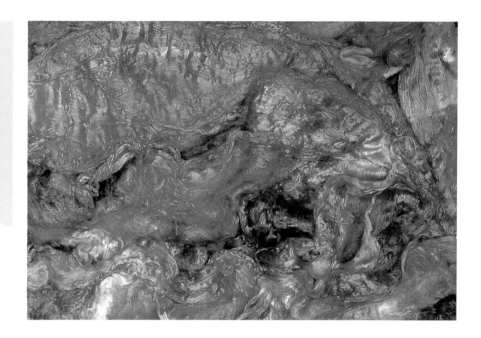

To make the cover

Reinforce with vilene then cut to size A4

1 Follow the basic silkpaper making technique to lay out two layers of turquoise silk fibre (see pages 64–67), to make an A4 pieceof paper.

2 Position the two lime-green feathers so that, when the paper is folded in half, one will appear on the front of the book and the other on the back. Take small sections of turquoise silk fibre and lay them over the top and bottom of each feather, so that they are trapped in place.

3 Follow the basic silkpaper making techniques to glue, dry and remove from net (see pages 64–7). Strengthen with iron-on vilene.

Materials and Equipment

Binding:

Basic single-section binding equipment, plus

Cover: one A4 sheet of silkpaper with feathers, strengthened with vilene

Pages: three sheets of lime A4 handmade paper

Binding: wide, two-tone ribbon in blue/green

Method

Follow the single-section method of binding (see pages 129–131).

A4 silkpaper

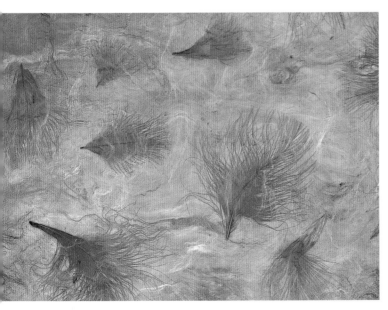

Floating pulp (cloud paper)

Project:

An A5 single-section book with a floating pulp (cloud paper) cover design

Inspiration:

Rock pools

Rock pools can be very mysterious, with their hidden depths and floating seaweeds obscuring your view – a palette of soft pastel shades floating randomly in the water above smooth, sea-worn stones. Use these colours to create your own pools of mystery with floating pulp.

Materials and Equipment

Cover:

Basic pulp-papermaking equipment, plus

A4 mould and deckle

Selection of green and blue pulp

One sheet of pale green A4 handmade paper

To make the cover

1 Pour two different colours of pulp into separate containers.

2 Float the mould and deckle in a vat that is three-quarters full of water. Agitate to get rid of any air bubbles.

3 While holding the mould and deckle steady, pour on a little of the coloured pulp, one colour at a time. You should still be able to see the mould mesh.

4 Use your fingers to move the floating colours around and distribute them as you wish.

5 Lift the mould and deckle out of the vat, sponge off excess water from underneath the mould, then remove the deckle.

Materials and Equipment

Binding:

Basic single-section binding equipment, plus

Cover: one A4 sheet of green/blue cloud paper

Pages: three sheets of A4 handmade pale green paper

Binding: wide blue and white gingham ribbon

6 Take the sheet of pale green A4 paper and place it on top of the mould. This will hold the blobs of coloured pulp together, as they would not form a sheet on their own. The dry sheet of paper will attach itself easily to the blobs of coloured pulp, so there is no chance of it falling off.

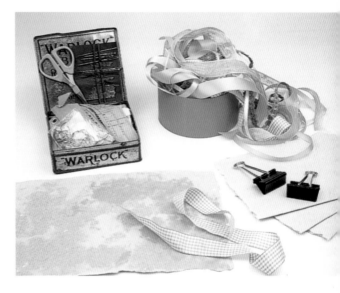

7 Couch, press and dry, following basic papermaking techniques (see page 43-47).

Method

Follow the single-section method of binding (see pages 129–131).

Burnt silkpaper

Project: An A5 single-section book with a burnt silkpaper cover

Inspiration: Fish skeleton bones

Fish bones are very striking, but with a head and tail still
attached they are much more interesting. The colours are
not the most exciting, so I have gone over the top with
brilliant red and dark blue, burning through silkpaper to
make sharp, stencil-like designs. Shine a light through your
paper and see the fish shadow come alive.

Materials and Equipment

Cover:

Basic silkpaper making
  equipment, plus

Red silk fibre

Iron-on vilene

Pyrography machine

Wooden board

Pencil

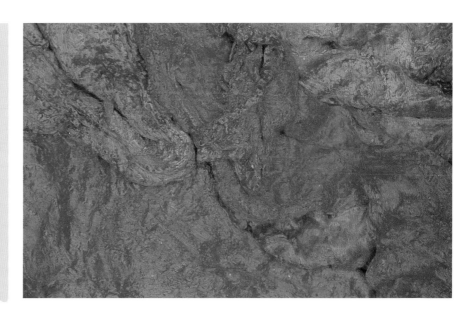

To make the cover

1 Use the red silk fibre to make a sheet of A4 silkpaper, following the basic silkpaper making techniques (see page 64–7); glue, dry, and remove from the net.

2 Strengthen the silkpaper with iron-on vilene then, using the pyrography machine, burn the paper to A4-size. The picture, bottom right, shows the burnt edge.

3 Draw stencil-like fish skeleton designs on half of the sheet, on the vilene side.

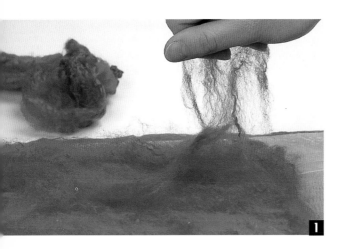

1

1

2

Materials and Equipment

Binding:

Basic single-section binding equipment, plus

Cover: One A4 sheet of vilene-strengthened red silkpaper with burnt fish skeleton designs

Pages: Three sheets of dark blue, A4 handmade pulp paper

Binding: Light blue cord

A4 Silkpaper

4 Working on a board to protect your work surface, follow the lines carefully, burning through the paper with the pyrography machine.

Method

Bind the book, following the single-section method of binding (see pages 129–131).

Paste paper

Project:

An A5 single section book with a paste-paper design cover

Inspiration:

Crashing waves

Rolling, crashing waves dance along the shore and sun shining through the crest of the waves illuminates the water, turning it many different shades of blue. Scraping through thick, dark-blue-dyed flour paste will also reveal patterns of lighter blue. Use a simple comb to make a bold statement with flowing lines and curves.

Materials and Equipment

Cover:

One sheet of white A4 handmade pulp paper

White flour (any sort)

Wooden spoon

Teaspoons

Measuring jug

Weighing scales

Hotplate

Small saucepan

Dark blue direct dye, or powder paint

A6 stiff plastic

Scissors

Wide paintbrush

Protective plastic sheet

Rubber gloves

Kitchen cloth

This works best with non-absorbent paper. Use spray-on starch to add extra size if necessary.

To make the cover

1 First make the flour-and-water paste. To do this, mix 40gms (1½ oz) of flour and 500ml (1 pint) of water together in a small pan.

2 Leave for 30 minutes, or until any lumps have completely dissolved.

3 Slowly bring to boiling point, stirring all the time, then simmer for 10 minutes.

4 Pour into a bowl and leave to cool. It should be the same consistency as custard. If it is too thick, gradually add water, stirring all the time until it is the right consistency. There is no need to reheat. If kept in a refrigerator, the paste will keep for a few days.

5 Wearing rubber gloves, add 1gm (½ teaspoon) of either blue direct dye, powder paint, or pigment, and mix to make a dark-blue paste.

6 Cut a small simple comb out of the stiff plastic, by making very small 'v' cuts with the points of sharp scissors.

7 Cover the work surface with protective plastic. Spread out a kitchen cloth over it, then place the paper on top. Pour out some of the coloured paste along one edge.

Tip

Flour-and-water paste can also be used as glue that is reversible when remoistened

8 Spread out the paste evenly, with the wide paintbrush.

9 Holding the plastic comb firmly, draw simple wave patterns. At the end of each row, remove the coloured paste that has built up on the plastic comb by wiping it off on the side of the bowl.

10 Continue until the whole sheet is patterned.

11 Keep the sheet flat, and leave to dry overnight on the cloth. Once dry, the cover design is ready to use.

Method

Follow the single-section binding method (pages 129–31).

Materials and Equipment

Binding:

Basic single-section binding equipment, plus

Cover: one A4 sheet of blue paste paper

Pages: three sheets of blue, handmade A4 paper

Binding: wide denim tape

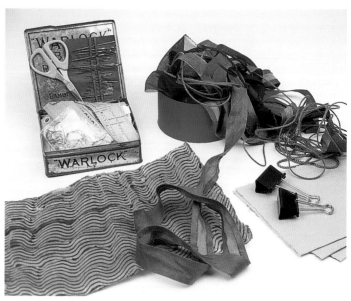

Single-section binding gallery

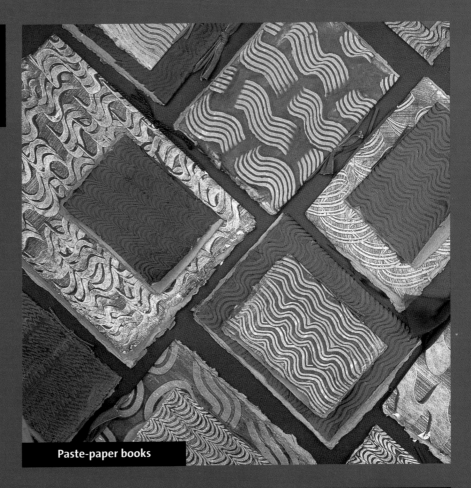

Paste-paper books

Floating-pulp-paper books

Burnt-silkpaper book

Burnt-silkpaper books

Burnt silkpaper

Silkpaper books with feathers

Floating-pulp-paper books

Paste-paper books

Multi-section Binding

This will make a chunky book that will be admired. Don't be daunted by all the measurements and the number of sewing needles, just get stuck in. The rewards will be great if you take your time and follow the step-by-step instructions.

Tough linen tape will make a strong binding for a multi-section book, but you can also use decorative alternatives to the tape, such as florists' paper ribbons or wide strips of patterned cotton fabric, folded to size. Whatever you choose, it should be both strong and flexible; and always use thick linen thread for the stitches, as thin threads will tend to cut through handmade paper.

Here, extra strength has been given to the spine of the book by interleaving two sheets of tissue paper made from abaca (a type of banana plant), which is very thin and strong. The three layers are joined together during the process of papermaking, so no glue is needed. Thin cotton fabric could be used as an alternative, but experiment with other materials and test the strength of the resulting paper before you bind it into a book.

Materials and Equipment

Two A5 sheets of lime-green handmade paper, with hidden spine-reinforcement strip (for the cover)

18 x A5 sheets of blue handmade paper with hidden spine-reinforcement strip (for the pages)

2cm (¾in) wide linen tape: two pieces, each 15cm (5⅞in) long

Thick linen thread

Five sewing needles

Beeswax

One large nail

Hammer

Wooden board

Scissors

Pen

Ruler

Bone folder

Glue

I have split open this dry sheet of paper to reveal the strip of abaca tissue paper

A5 sheets of paper will make an A6 multi-section book.

For each sheet of paper with reinforced spine:

● Make a sheet of A5 paper.

● Cut strips of abaca tissue paper 2.5cm wide x 13.5cm long (1 x 5¼in), and place a strip along the spine edge of your paper while the paper is still wet. The tissue will be a little shorter than the spine of your book, so that it is completely concealed.

● Make a second sheet of A5 paper and couch it on top of the first sheet, so that the tissue paper is completely sealed inside and hidden.

● Follow the basic pulp-papermaking process for pressing and drying the layered sheet (see pages 45–7)

1 Fold the sheets in half, carefully aligning the edges of each one, then use the bone folder to press down on the fold to get a sharp, smooth crease. This folded edge is the spine of your book and will be stitched along.

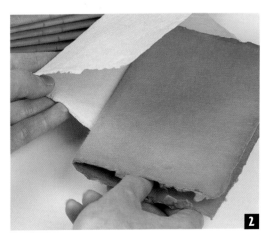

2 Each section is made from two folded sheets of strengthened paper, one placed inside the other, so these 20 sheets of paper will give you ten sections, including the cover.

For both the back and front covers, place one sheet of folded blue paper inside a folded sheet of lime-green paper. For each of the eight centre sections, use two sheets of blue paper, one folded inside the other. Stack the ten sections up in the order they will be stitched, i.e. with the lime-green covers at the bottom and top of the stack.

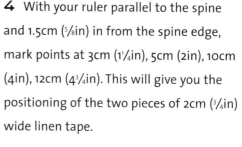

3 On both lime-green covers, mark a line 1.5cm (⅝in) in from the spine edge. This is where the linen tape will be placed.

4 With your ruler parallel to the spine and 1.5cm (⅝in) in from the spine edge, mark points at 3cm (1¼in), 5cm (2in), 10cm (4in), 12cm (4¾in). This will give you the positioning of the two pieces of 2cm (¾in) wide linen tape.

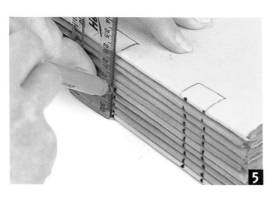

5 Use a ruler against the spine of your book to mark the stitch lines. These should be just slightly in from the lines drawn on the top cover, which indicated the exact width of the linen tapes, as the stitching will of course go through the linen tape, rather than to the side of it. There are two rows of stitching for each linen tape, so a total of four rows to be marked.

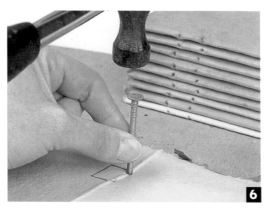

6 Open out each section and strike a hole at each mark with the hammer and nail, then replace the sections in the same order, ready for stitching.

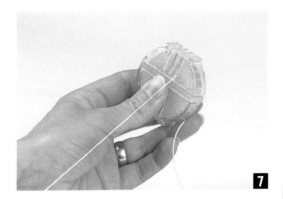

Thread Lengths

You will need three separate pieces of linen thread: the first will be 30cm (11¾in) long, for decorative stitching, and is used with one sewing needle. The other two pieces of linen thread are used for binding the pages of your book together, one thread for each linen tape.

To calculate the lengths of thread required, multiply the number of sections in your book by twice the width of the linen tape, then add 50cm (20in) for ease of stitching, and to allow for thick pages.

So, for this book you will need two lengths of thread 10 times 2 x 2cm (= 40cm/15¾in) + 50cm (20in) = 90cm (35¾in) long, one for each linen tape.

The threads are used with a sewing needle attached to each end, so you need a total of four sewing needles, which will be used in pairs.

7 Wax all the threads, to help prevent tangles and to stop the thread slipping once it has been pulled tight.

8 Attach all of the needles in this way: thread the needle and pull a short tail through. Split the short tail with your needle. Hold the pointed end of the needle then pull the longer length of thread.

9 Fold in a small hem at one end of each linen tape. Use the short length of thread, with one sewing needle, to make a decorative running stitch and hold the hem in place. The stitched areas should be the same size as those marked out on the lime-green cover.

10 Glue both linen tapes in place with a fabric glue.

Note

Always keep the book the same way up while stitching; the decorative stitching on the front cover should be facing the table top at all times.

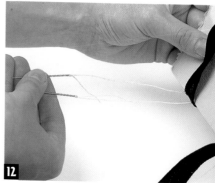

11 Next, attach the linen tapes one at a time: first, take one of the 90cm (35¾in) lengths of waxed linen thread, with a sewing needle attached to each end. Open the first section to the centre page. Thread the first pair of needles through the stitch holes for one of the linen tapes, starting from the centre of the section, then through the linen tape to the outside of the book.

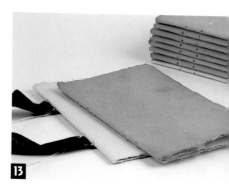

12 Hold both needles and pull tight so they are the same length. Repeat steps 11 and 12 for the other linen tape, using the second pair of needles.

13 Add the second section.

14 Open it out to the centre page. Make a running stitch in the linen tape and push the first pair of needles through the stitch holes in the next section.

15 Push both needles through to the centre of the section and pull tight.

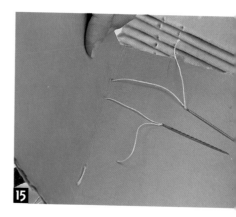

16 Crossing the threads over, take the first pair of needles back through the opposite stitch holes and out through the same linen tape.

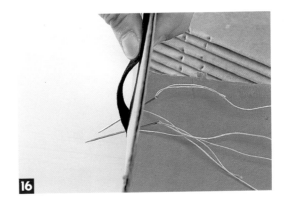

17 Holding the sections closed, pull the threads parallel to the spine until tight. Repeat steps 14, 15, 16 and 17 for the other linen tape and the second pair of needles. Then repeat steps 13, 14, 15, 16 and 17 until all ten sections are bound together.

18 Finish off with decorative stitching. Start with the first pair of needles. Take the needle nearest the side of the book, and make a running stitch through the tape only.

19 Take this needle, cross over the top of the linen tape, push the needle through the inner stitch hole, then through to the centre of the last section.

20 Trim any excess linen tape and fold in, to fit the area marked on the cover. With the remaining needle, hold the folded linen in place with a running stitch. You should end at the stitch hole nearest the side of the book.

21 Push the needle through the linen tape and into the centre of the last section.

22 Repeat steps 18–21 for the other linen tape and second pair of needles. All four needles should now be inside the final section.

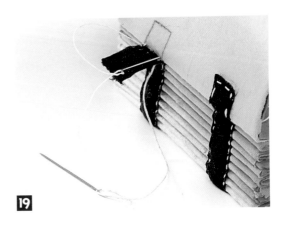

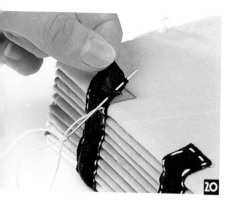

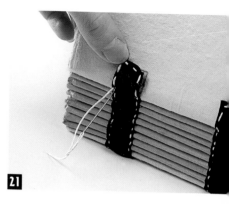

23 Cut all needles off the thread. Tie the thread ends into two bows.

24 Glue the decoratively stitched linen-tape ends in place.

25 Your multi-sectioned book is now complete.

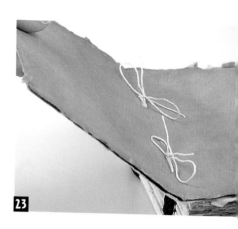

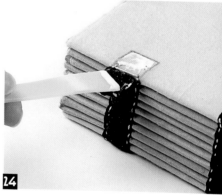

Ripping through wet pulp

Project:

A6 multi-section book with a torn-paper cover

Inspiration:

Rocks

Deep crevices in rocks, bottomless gaps between boulders, evidence of many harsh winters. Make your papers sculptural by forming gashes in newly formed sheets of wet paper, carefully manipulating the torn edges to make the gullies deeper and wider.

Materials and Equipment

Cover:

Basic pulp-papermaking equipment, plus

A6 mould and deckle

Dark blue pulp

Lime-green pulp

One sheet of dark blue A5 handmade paper with hidden spine-reinforcement strip

One sheet of lime green A5 handmade paper with hidden spine-reinforcement strip

Strong thread

Scissors

To make the cover

1 Lay the sheet of dark blue handmade paper with hidden spine-reinforcement strip onto the kitchen cloth. Sprinkle, or squeeze a sponge full of water over it, to dampen. Do not touch the paper itself, as it could easily tear and any wrinkles present will be removed when the second layer is couched on top. Place a length of strong thread parallel to the width, and one quarter in from one edge.

2 Place the sheet of A6 lime-green paper on top of the thread. Press with a sponge, then remove the mould.

3 Carefully remove the thread by tearing it through the lime-green paper.

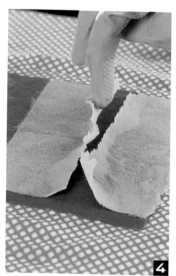

4 Use your fingers to open the gap wider. Do not cover. Leave on the kitchen cloth and dry on a flat surface. This will keep the torn paper more sculptural.

5 Repeat the process for the back cover, but reverse the colours.

Method

Follow the multi-section method of binding (see pages 153–9).

Materials and Equipment

Binding:

Basic multi-section binding equipment, plus

Cover: two sheets of A5 torn paper with rock design, each with hidden spine-reinforcement strips (for the cover)

Pages: 18 sheets of A5 handmade blue and lime-green paper, each with hidden spine-reinforcement strips

Binding: dark blue linen tape and green linen thread

The completed book

Embossing found objects

Project: An A6 multi-section book with an embossed driftwood cover

Inspiration: Driftwood

The raised grain of sandblasted, weather-worn driftwood can make attractive covers for a book. However, you may not come across any driftwood that is wide enough, and to make a matching front and back cover could be even harder, as you could be beachcombing for quite some time. Instead, get the driftwood effect on your paper by using it for embossing, and dry your sheets of paper on the wood instead of kitchen cloths. You can use the same piece of wood to make many embossed papers. Some driftwoods will give up interesting colours from rusty nails and so on, so experiment with white pulp first, as these delicate marks may not show up on brightly coloured papers.

Materials and Equipment

Cover:

Basic pulp-papermaking equipment, plus

A5 mould and deckle

White pulp

Spine-reinforcement strips

Coffee-coloured direct dye

Jug

Spoon

Sponge

Plastic gloves

Protective plastic sheet

Driftwood with a textured surface

A6 thick card or plastic

To make the cover

1 Mix a strong solution of coffee-coloured direct dye and water then, wearing gloves, sponge onto one side of the driftwood. The dye will help emphasize the grain of the wood, picking up small details. The coffee colour will give your paper the appearance of wood.

2 You will only require the embossed driftwood pattern on half of the A5 paper, as the other half will be folded inside to become one of the pages. To keep your paper flat and to protect the other half from the wood dye, lay out a wood support, covered with a piece of card or plastic.

3 Press firmly with a sponge.

4 Lay out a spine-reinforcement strip along the centre, where the fold of the spine will be.

5 Cover with a second sheet of white A5 paper. Press firmly with a sponge, then remove the mould.

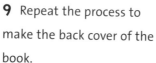

6 Cover with a protective kitchen cloth.

7 Sponge firmly all over, to ensure that the paper picks up the wood-grain texture.

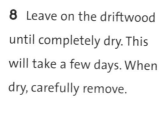

8 Leave on the driftwood until completely dry. This will take a few days. When dry, carefully remove.

9 Repeat the process to make the back cover of the book.

Method

Follow the multi-section method of binding (see pages 153–9).

Materials and Equipment

Binding:

Basic multi-section binding equipment, plus

Cover: two sheets of A5 embossed-driftwood paper, each with hidden spine reinforcement strips

Pages: 18 sheets of A5 handmade white paper each with hidden spine reinforcement strips

Binding: natural linen tape and orange linen thread

The finished embossed driftwood book

Embossing your own designs

Project:

An A6 multi-section book with an embossed shell cover

Inspiration:

Scallop shells

The scallop shell has provided me with a ready-made symmetrical design that can be drawn onto printing lino. You only need to cut one design, as it can be reused for both the back and front covers of the book. If you creatively position the embossed scallop shells with hinges nearest the spine of the book, you can make it look like a whole shell, front and back.

Materials and Equipment

Cover:

Basic pulp-papermaking equipment, plus

A5 mould and deckle

Red pulp

Yellow pulp

Spine-reinforcement strips

Printing lino 22 x 30cm (8⅝ x 11¾in)

Linocutters

Pencil

Iron

To make the cover

1 Use the A5 deckle to draw guidelines on the lino (the shaded area shown, below right, represents the deckle). Divide into two A6 areas. On one side mark out the positioning of the 2cm (¾in) wide linen tape. This will be 1.5cm (⅝in) in from the centre line (spine edge). Mark points at 3cm (1⅛in), 5cm (2in), 10cm (4cm) and 12cm (4¾in).

Draw a simple symmetrical scallop shell with the shell hinge nearest the centre line (spine of the book).

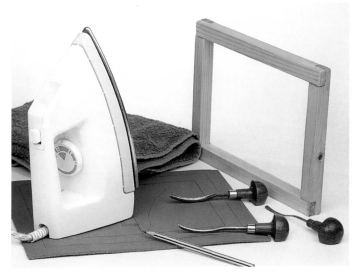

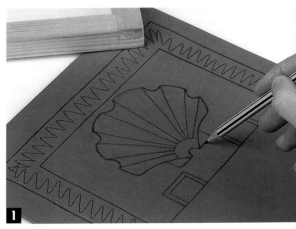

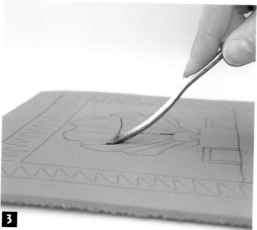

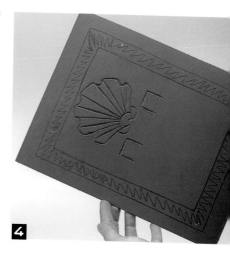

2 Iron the lino, to make it soft and much easier to cut.

3 With a 'V' shaped lino-cutter, follow the lines of the shell carefully, cutting away from your body and keeping all fingers out of the way. Also cut out the 'U' shapes which mark where the linen tapes will be positioned. If the lino gets cool and hard to cut, apply more heat with the iron.

4 The shell linocut is now ready to use.

5 Make a sheet of A5 red paper, using the shaded guide lines on the lino to position it.

6 Press firmly with a sponge, then remove the mould.

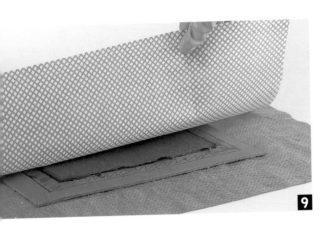

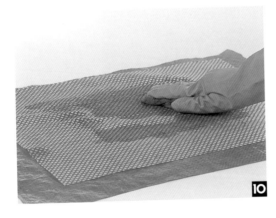

7   Place a strip of tissue paper along the centre, where the spine fold will be.

8   Cover with a second sheet of red A5 paper.

9   Cover with a protective kitchen cloth.

10   Sponge firmly all over to make sure the paper fills all of the relief shell pattern.

11   Leave on the lino for a few days, until completely dry. Once dry, remove it carefully.

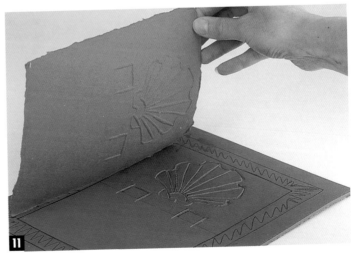

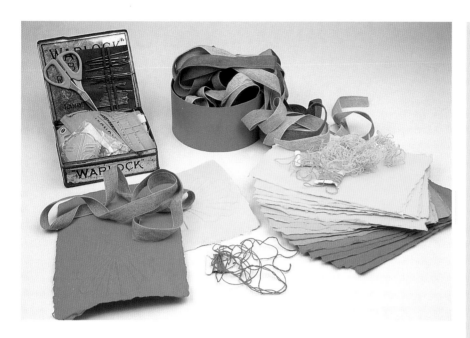

Materials and Equipment

Binding:

Basic multi-section binding equipment, plus

Cover: 2 sheets of A5 embossed shell paper, each with hidden spine-reinforcement strips

Pages: 18 sheets of A5 handmade red/orange/yellow paper, each with hidden spine-reinforcement strips

Binding: orange linen tape and red linen thread

12 Using the same shell-design linocut, repeat the process with yellow pulp to make the back cover of your book.

Method

Follow the multi-section method of binding (see pages 153-9).

The finished embossed shell book

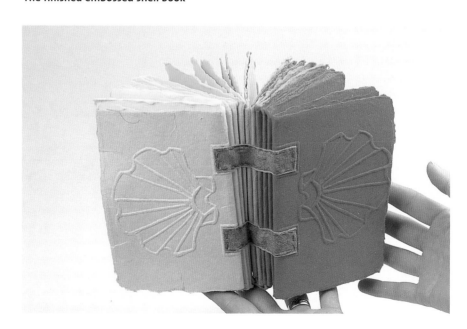

Computer-printed images

Project:

An A6 multi-section book with a computer-printed image cover

Inspiration:

Fishy family photographs

No scrapbook would be complete without family photos. I've selected photos from my own family album – all with fish. All you need is access to a computer with a scanner and a colour printer. This could make a very personalised gift for someone special, without even marking the original photo.

Materials and Equipment

Cover:

4 sheets of A5 white paper

2 spine-reinforcement strips

Scissors

Spray-mount adhesive

Family photos

Computer

Scanner

Colour printer

To make the cover

I found that the computer printer I used only accepted single-thickness sheets of paper. As all printers are different, be prepared to experiment.

1 First, scan a photograph onto your computer. Then, to make the back cover, print out onto the left half of an A5 sheet of single-thickness white handmade paper. To make the front cover, print out onto the right half of an A5 sheet of single-thickness white handmade paper.

2 On a blank single-thickness sheet of white A5 handmade paper, lay out a strip of the tissue paper along the centre, where the spine fold will be.

3 Making sure you are working in a well-ventilated area, apply a spray adhesive to the back of your printed paper.

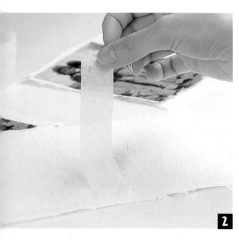

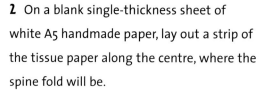

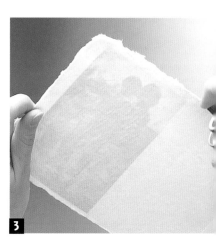

4 Press the three layers firmly together.

Repeat steps 2, 3 and 4 to make the back cover.

Method

Follow the multi-section method of binding (see pages 153-9).

Materials and Equipment

Binding:

Basic multi-section binding equipment +

Cover: 2 sheets of A5 printed photograph paper each with hidden spine reinforcement strips

Pages: 18 sheets of A5 handmade white/blue paper, each with hidden spine reinforcement strips

Binding: dark blue linen tape and natural linen thread

The finished book with printed photographs on the cover

Multi-section binding gallery

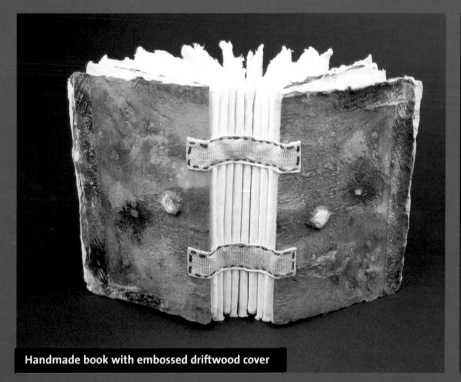

Handmade book with embossed driftwood cover

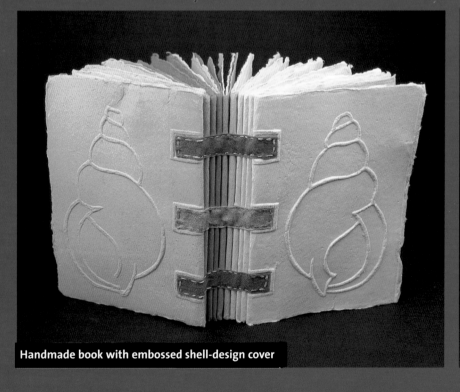

Handmade book with embossed shell-design cover

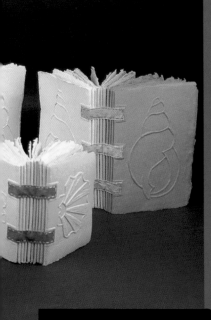

Handmade books with embossed shell-design covers

Handmade book with torn wet-pulp cover

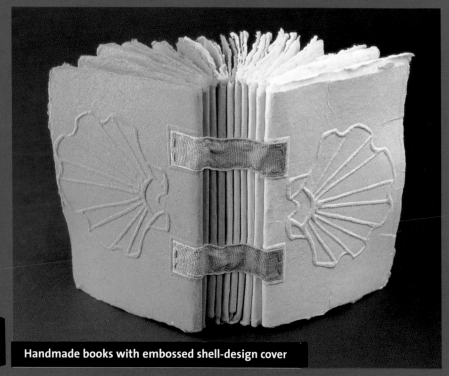

Handmade books with embossed driftwood covers

Handmade books with embossed shell-design cover

Glossary

Anionic A negative charge.

Blender Kitchen equipment used to make pulp from linters and wastepaper. Uses a cutting action so it makes shorter fibres.

Cationic A positive charge.

Cellulose Inner cell wall of plants used to make paper.

Cloud paper Decorative technique, floating different colours of pulp on water.

Couching Action of removing newly formed sheet of paper from mould to felts or kitchen cloths.

Deckle edge Feathered, uneven edge of handmade paper created during sheet formation, caused by the pulp escaping under the deckle.

Deckle frame This sits on top of a mould during sheet formation to hold in pulp and also to determine the size of the paper being made.

Direct dye Colouring which has a direct affinity to natural fibre, but requires salt or vinegar to improve colour take-up. Direct dye requires a rinse agent, as it is not wash-fast when used with half-stuff.

Embossing Relief image in paper formed when it has been pressed into, and left to dry on, a textured surface.

European mould A frame with a mesh fixed to it. Used in the process of papermaking, it acts like a sieve and is used in conjunction with a deckle to hold in the pulp.

Felts Used to couch papers on. Kitchen cloths are a substitute.

Half-stuff Plant fibre which has been processed and is halfway to being made into paper. It is sold in concentrated form.

Hollander beater Industrial-size machinery which processes plant material for papermaking by separating and flattening the fibre, resulting in fibres that have a larger surface area and are still long. This process is called macerating.

Inclusions Decorative technique – adding leaves, petals, glitter, etc. to the vat of pulp before a sheet of paper is formed.

Keta Oriental, hinged papermaking-frame.

Landscape paper Decorative technique, using more than one colour of pulp and overlapping half-sheets of paper to build up the layers of a landscape.

Lightfastness A test to see how much the colour in paper will fade when left in a variety of light conditions.

Linters The waste fibre left over from cotton when processed in a linting machine. Sold in concentrated form for papermaking.

Lye Used to accelerate rotting process.

Macerate The process of beating fibre by flattening and separating it.

Neri A formation aid used in Oriental papermaking which slows down drainage of water through the sugeta (see below) by thickening it. Neri also helps to distribute the long plant fibres evenly. It can be made from many different plants including the pod of the okra.

Nepalese/Indian deckle box High-sided deckle used on top of a mould and placed in a vat of water. Enough pulp for one sheet is poured into the deckle box before being lifted out of the water.

Paste paper A decorative technique, using flour and water which has been cooked to form a thick paste and mixed with strong colours. This is brushed on paper then patterns are combed through the paste while it is still wet.

pH A scale of 1 to 14 on which the acidity or alkalinity of a solution is measured. pH7 is neutral, below this is acid, and above is alkaline.

Pigment Coloured rocks which have been ground up. They are very lightfast.

Post Alternate layers of handmade papers and felts or kitchen cloths.

Pulp Plant fibre which has been processed into a sludgelike texture for papermaking.

Rain paper A decorative technique: droplets of water are splashed onto a newly formed sheet of paper while it is still on the mould, making a random perforated pattern.

Retention agent Added to give fibre a positive charge (cationic) when colouring with pigments.

Rinse aid Added at the end of the dyeing process, to help reduce colour-stain in the water and also reduce the amount of rinsing required. Improves colour take-up.

Size A waterproofing substance, such as gelatine or Aquapel, that reduces the absorbency of paper. It can be used in two ways, internal sizing or external sizing. For internal sizing, size is added to the pulp while it is in the vat, before a sheet of paper is made. For external sizing, individual sheets of paper are coated with size.

Spine The edge of a book which has been bound or stitched.

Sugeta Oriental papermaking frame: a separate flexible bamboo mat (su) is trapped in a hinged deckle frame (keta). Used together they are called a sugeta. Paper is formed by layers of long plant fibres, gradually built up using a repeated wave action when the sugeta is dipped in the vat, and the excess pulp and water are thrown off the top.

Vat Container which holds water and pulp. It must be big enough to accommodate the mould and deckle.

Washfastness A test to see how much colour will run out of fibre when it is rinsed or washed.

Watercut Decorative technique, drawn with a jet of water from a squeezy bottle onto a newly formed sheet of paper while it is still on the mould, making a perforated pattern.

Waterleaf Paper that has not been sized (waterproofed). If using inks they will bleed into the paper leaving a pattern like the veins in a leaf.

Watermark Made with a wire design which is stitched to the mesh on a papermaking mould. The paper pulp will be displaced by the wire, leaving the resulting papers thinner where the wire has been. The design is only visible when paper is held up to the light.

Wet lap Wet lap is the same as half-stuff, except that it is sold in wet form, in a plastic bag. The reason for this is that each time the plant fibre is dried then rehydrated, the cellulose will take in less and less water, resulting in weaker paper. If you are not going to use it straightaway, you can freeze it for use later – this goes for any wet pulp that you would rather not dry out. However, it must be labelled clearly and packaged away from food.

Wetting agent Added to fibre to help it to accept dye, and allow it to penetrate.

Useful Addresses

Australia

Primrose Paperworks Co-operative
(Cremorne)
c/o 107 Hunter Avenue
St Ives
NSW 2075
Australia

Tel: 02 9546 6558/02 9949 5607
Email: cdrew@ihug.com.au
Website: www.primrose-
park.com.au/paper/papermaking.htm

Open-access papermaking studio.

Southern Papermakers
Margaret Mason
4 Tiranna Place
Oyster Bay
NSW 2225
Australia

Tel: 02 9528 5008
Email: kozo63@netspace.net.au

Organisation with a special interest in
papermaking and associated techniques.

Treetops Colour Harmonies
Nancy Ballesteros
6 Benween Road
Floreat 6014
Western Australia

Tel: 08 9387 3007
Fax: 08 9387 1747
Email: nancy@treetopscolours.com.au
Website: www.treetopscolours.com.au

Hand-dyed silk fibres.

R K Graham
4 Tiranna Place
Oyster Bay
NSW 2225
Australia

Tel: 02 9528 5008
Email: rodmar@netspace.net.au

Pulp-papermaking equipment: moulds
and deckles size A6 to A3, also deckle
boxes and envelope deckles.

Canada

Canadian Bookbinders and Book Artists
Guild (CBBAG)
176 John Street
Suite 309
Toronto, Ontario
Canada
M5T 1X5

Tel: 416 581 1071
Fax: 905 851 6029
Email: cbbag@web.net
Website: www.cbbag.ca

Handmade book-arts guild.

The Japanese Paper Place
887 Queen Street West
Toronto, Ontario
Canada
M6J 1G5

Tel: 416 703 0089
Fax: 416 703 0163
Email: washi@japanesepaperplace.com
Website: www.japanesepaperplace.com

All bookbinding materials, equipment and
handmade papers.

The Papertrail
170 University Avenue West
Suite 12-214
Waterloo,
Ontario, Canada
N2L 3E9

Tel: 519 884 7123 or
1 800 421 6826 toll-free in North America
Fax: 519 884 9655
Email: info@papertrail.ca
Website: www.papertrail.ca

All materials and equipment for pulp-
papermaking and bookbinding.

Treenway Silks
501 Musgrave Road
Salt Spring Island
BC V8K 1V5
Canada

Tel: 250 653 2345 or
1 888 383 745 toll free
Email: silk@treenwaysilks.com
Website: www.treenwaysilks.com

All silkpaper making materials and
equipment.

Germany

IAPMA (International Association of Hand
Papermakers and Paper Artists)
Eva Maria Juras (IAPMA secretary)
Tulpenstrasse 20
51427 Bergisch Gladbach
Germany

Tel: 09 2204 678 72
Fax: 09 2204 961 428
Email: evajuras@aol.com
Website: www.iapma.info

An international organisation with a
special interest in papermaking, paper
arts and bookbinding.

Eifeltor Mühle
John Gerard
Auf dem Essig 3
D-53359 Rheinbach
Germany

Tel: 0 2226 2102
Fax: 0 2226 913 437
Email: gerard@eifeltor-muehle.de
Website: www.eifeltor-muehle.de

All materials and equipment for pulp
papermaking.

Taiwan ROC

Goang Horng Shing Hand Made Paper
Manager: Huann-Jang Hwang
No. 310 Teih-sang Road
Pu-Li Chen
Nan-Tou County
Taiwan
ROC

Tel: 49 2913037 or 2916899
Fax: 49 2913038
Email: paper@taiwanpaper.com.tw
Website: www.taiwanpaper.com.tw

Handmade paper.

UK

The Handweavers Studio
29 Haroldstone Road
Walthamstow
London
E17 5AN

Tel: 020 8521 2281
E-mail: handweaversstudio@msn.con
Website: www.handweaversstudio.co.uk

Silk fibre and dyes.

HT Gaddum & Co Ltd
3 Jordangate
Macclesfield
Cheshire
SK10 1EF

Tel: 01625 427666
Fax: 01625 511331
Email: merchanting@gaddum.co.uk
Website: www.gaddum.co.uk

Silk fibre.

Homecrafts Direct
PO Box 38
Leicester
LE1 9BU

Tel: 0116 2697733
Fax: 0116 2697744
Email: info@homecrafts.co.uk
Websites: www.homecrafts.co.uk
www.studentsartshop.co.uk

All materials and equipment for art and
crafts including pulp-papermaking and
bookbinding.

John Purcell Paper
15 Rumsey Road
London
SW9 OTR

Tel: 020 7737 5199
Fax: 020 7737 6765
Email: jpp@johnpurcell.net
Website: www.johnpurcell.net

Linters, half-stuff, pigments and
handmade papers.

Kemtex Educational Supplies
Chorley Business & Technology Centre
Euxton Lane
Chorley
Lancashire
PR7 6TE

Tel: 01257 230220
Fax: 01257 230225
Websites: www.kemtex.co.uk
www.textiledyes.co.uk

All materials and equipment for dyes,
including direct dyes.

The Paper Shed
Tollerton
York
YO61 1QQ

Tel: 01347 838256
Fax: 01347 838253
Email: papershed@papershed.com
Website: www.papershed.com

All materials and equipment for pulp and
silkpaper making; also stock handmade
papers.

Paperweight
Website: www.paperweight.demon.co.uk

UK group which focuses on paper as an
art form.

The Scottish Bamboo Nursery
John Wojciechwski
Middlemuir Farm
Craigievar
Alford
Aberdeenshire
AB33 8JS

Tel: 019755 81316
Fax: 019755 81411
Email:
info@thescottishbamboonursery.co.uk
Website: www.thescottishbamboonursery.
co.uk

Bamboo plants.

Shepherds Bookbinders
76 Rochester Row
London
SW1P 1JU

Tel: 020 7620 0060
Fax: 020 7931 0541
Email: shepherds@bookbinding.co.uk
Website: http://www.bookbinding.co.uk/

All materials and equipment for
bookbinding.

George Weil & Sons Ltd
Fibrecrafts
Old Portsmouth Road
Guildford
Surrey
GU3 1LZ

Tel: 01483 565800
Fax: 01483 565807
Email: sales@georgeweil.co.uk
Website: www.fibrecrafts.com

All materials and equipment for pulp and
silkpaper making, also dyes and related
equipment.

USA

Carriage House Paper
76 Guernsey Street
Brooklyn
New York 11222
USA

Tel/Fax: 718 599 PULP (7857)
Email: chpaper@aol.com
Website: www.carriagehousepaper.com

All materials and equipment for pulp-
papermaking.

Lee S. McDonald Inc
PO Box 290264
Charlestown
MA 02129
USA

Tel: 888 627 2737 (toll-free in US) or 617 242
2505
Fax: 617 242 8825
Email: mcpaper@aol.com
Website: www.toolsforpaper.com

All materials and equipment for pulp-
papermaking.

Twinrocker
Mailing address:
Kathryn Clark
Twinrocker Handmade Paper
PO Box 413
Brookston
IN. 47923
USA

Location:
100 East Third Street
Brookston
IN. 47923
USA

Tel: 765 563 3119
Fax: 765 563 8946
Email: twinrocker@twinrocker.com
Website: www.twinrocker.com

All materials and equipment for pulp-
papermaking and bookbinding.

Further Reading

Bell, Lilian A.
Plant Fibers for Papermaking
(Liliaceae Press,1981)

Hunter, Dard
*Papermaking: The History and Technique of
an Ancient Craft*
(Dover, 1947)
ISBN 0 486 23619 6

Ikegami, Kojiro
Japanese Bookbinding
(Weatherhill, 1979, English edition – 1986)
ISBN 0 8348 0196 5

Diehl, Edith
Bookbinding: Its Background and Technique
Two volumes bound as one.
(Dover 1980, from original volumes
published in 1946)
ISBN 0 486 24020 7

Index

abaca
  half-stuff 10
  tissue paper 153, 154
adhesives
  silkpaper making 62,
    65–6
  sizing see size
  waterproof wood glue
    29, 30
alkali 7
aluminium mesh 27, 28,
  30–1
amate 18
anchor design block
  printing 90–3, 96, 97
anionic charge 52, 180
Aquapel 25, 38

bamboo half-stuff 10–11, 21
bark 18
Bashania fargessi 10, 11
baster, turkey 25
beating see macerating
blender 8, 22, 36–7, 51, 54,
  80, 180
block printing 90–4, 96, 97
bone folder 130, 154
bowls 24, 63
Broughton, Jenny 1
bucket 23
burnt silkpaper 140–3,
  150–1

calligraphy paper 10, 11
card template 130
cationic charge 52, 180
caustic soda 7, 8
cellulose 7, 52, 55, 180
cellulose paste 25, 38, 62,
  65–6
ceremonial paper for
  burning 11
charge 52, 180
China 11, 14
cloud paper (floating pulp)
  136–9, 149, 151, 180
colour samples 53, 56, 69
colouring pulp 48–59
  direct dyes 55–9
  lightfastness test 49
  pH test 49
  pigments 52–4

pre-coloured
  wastepaper 51–2
  safety 50
comb, plastic 147, 148
commercial envelope 32, 33
computer-printed images
  174–7
copper wire 115
Corallina officinalis 79
couching 43, 180
Cyperus papyrus 17

deckle box and mould
  14–15
deckle edge 19, 100, 180
deckle frame 15, 180
  envelope-shaped 32–3, 35
  see also mould and
    deckle
decorative handstitch
  122–5, 127
Dendrocalamus strictus 11
direct dyes 55–9, 180
  colouring silk fibre 68–73
dowelling 28, 30
driftwood 2
  embossing with 164–7,
    178–9
drip tray 23, 39, 45
drying 46–7
dulse 79, 80
dye bath 57–8
dyes 50
  direct dyes see direct dyes
  leftover solutions 72
  see also pigments

Egypt 17
embossing 180
  driftwood 164–7, 178–9
  shell design 168–73,
    178–9
envelope-shaped deckle
  32–3, 35
envelopes 76–97
  forming 77
  printmaking 90–4, 96, 97
  rain paper 86–9, 95, 97
  seaweed inclusions
    78–81, 95, 96–7
  wastepaper inclusions
    82–5, 96–7
equipment
  for pulp papermaking
    20–5

for silkpaper making 61,
  62–3
European mould 15, 180
  see also mould and
    deckle
Evers, Inge 19
external sizing 47, 180

feathers
  inlay 132–5, 150–1
  watermarks 113–16, 126
felts 180
  see also kitchen cloths
fibre 21
  selecting 6–11
  silk fibre see silk fibre
fish
  inclusions of fish design
    82–5, 96–7
  skeleton design 140–3,
    150–1
fishing feathers 133–5
flax half-stuff 10
floating pulp (cloud paper)
  136–9, 149, 151, 180
flour-and-water paste
  146–8
formation aid 15
  see also neri
four-hole stitch 100–3
  see also Japanese stab
    binding
frames, papermaking 12–15
  see also mould and
    deckle

gelatine size 25, 38–9
glass/clear plastic jar 25
glue see adhesives
grass kelp 79, 81
grasses 119, 120
green hairweed 79, 81

half-stuff 9–11, 34, 35, 36,
  180
hand towel 24, 63
handstitch, decorative
  122–5, 127
hemp half-stuff 10
Hollander beater 180

inclusions 180
  seaweed 78–81, 95, 96–7
  wastepaper 82–5, 96–7
Indian/Nepalese deckle

box 14–15, 180
indoor washing line 25, 46,
  47, 63, 66
ink roller 92–3, 94
internal sizing 38–9, 44,
  180
iron-on vilene 67

Japanese stab binding
  98–127
  decorative handstitch
    122–5, 127
  landscape paper 118–21,
    126–7
  silkpaper cover 104–7, 126
  technique 100–3
  watercut designs 108–11,
    126–7
  watermarks 112–17, 126, 127
jug, measuring 24, 63

keta 13, 180
kettle 25
kitchen cloths 22, 39, 43,
  45, 46, 47

landscape paper 118–21,
  126–7, 180
lightfastness 52, 55, 180
  testing 49
lignin 7
linen tape 153, 155, 156–9
linen thread, thick 153,
  156–9
lino cut 169–73
linters 9–11, 180

macerating 7, 8, 180
mackerel 105
mackerel feathers 133–5
maiden's hair 79
measuring jug 24, 63
mesh, aluminium 27, 28,
  30–1
montbretia 120
mould and deckle 15, 21
  making 27–33
  pulling a sheet of paper
    40–1, 42
multi-section bookbinding
  152–79
  computer-printed
    images 174–7
  embossed found objects
    164–7, 178–9

embossed shell design 168–73, 178–9
technique 153–9
torn wet pulp cover 160–3, 179

Nepalese/Indian deckle box 14–15, 180
neri 13, 180
nylon net 19, 27, 28
silkpaper making 62, 64, 65, 66

paintbrush 93
panel pins 28, 30
paperclips 25, 63
papyrus 16, 17
part-processed fibres 9–11
paste paper 144–8, 149, 151, 180
pegs 25, 63
pH 180
testing 49
photocopied image inclusions 82–5, 96–7
photographs, scanned 174–7
pigments 50, 52–4, 180
see also direct dyes
plant fibre, raw 7–8
plastic comb 147, 148
plastic fish boxes 123
plastic sheet 63, 64
polystyrene printing tiles 94
polythene food wrap 71–2
post 13, 15, 23, 180
couching 43
pressing 44–6
pre-coloured wastepaper 51–2
press, standing 23, 45
pressing
paper 44–6
seaweed 80
pressing boards 23
printmaking 90–4, 96, 97
pulling a sheet of paper 40–1
thickness problems 42
pulp 42, 180
adding pulp to the vat 44
colouring see colouring pulp
preparing 35–7

torn wet pulp covers 160–3, 179
pulp papermaking
technique 34–47
adding more pulp to the vat 44
adding size to the vat 38–9, 44
couching a post of paper 43
drying 46–7
external sizing 47
preparing the pulp 35–7
preparing a work surface 39
pressing the post 44–6
pulling a sheet of paper 40–1
surface texture 47
thickness problems 42
PVA white glue 25, 38
pyrography machine 142–3

raffia 123, 125
rain paper 86–9, 95, 97, 180
rainbow dyeing 71
raw plant fibre 7–8
recycled paper see wastepaper
retention agent 52, 53, 54, 180
retting 7, 8
ribbon 3, 130–1
rinse aid 56, 58, 180
rock pools 2, 137
ropes 123
rubber gloves 24, 63

safety 36, 50, 70
salt 56, 57
scallop shell design 168–73, 178–9
scalpel 91
scanned photographs 174–7
sea-fishing feathers 133–5
sea lettuce 79, 81
seashore 2
seaweed inclusions 78–81, 95, 96–7
shells 2
embossed shell design 168–73, 178–9
sieve 23
silk fibre 19, 62, 64–5

colouring with direct dye 68–73
silkpaper 19, 60–7
adding strength 67
burnt 140–3, 150–1
cover for Japanese stab book 104–7, 126
with decorative handstitching 122–5, 127
equipment for making 61, 62–3
with feather inlay 132–5, 150–1
techniques for making 64–6
single-section bookbinding 128–51
burnt silkpaper 140–3, 150–1
floating pulp 136–9, 149, 151
paste paper 144–8, 149, 151
silkpaper with feather inlay 132–5, 150–1
technique 129–31
size 25, 180
external sizing 47, 180
internal sizing 38–9, 44, 180
when to add more size to the vat 44
soap, liquid 70
soldering 114, 115
spine 100, 180
'spirit money' 11
sponges 24, 41, 63, 65
spoons 25, 63
spray-on starch 25, 47, 146
squeezy bottle 25
stab binding, Japanese see Japanese stab binding
standard paper measurements 27
standing press 23, 45
starch 25, 47, 146
steam-fixing dye 72
stitch-holes 100
stitch lines 155
strength 67
string 25, 63
sugeta 12, 13–14, 180
surface texture 47

tape, linen 153, 155, 156–9
template, card 130
temporary indoor washing line 25, 46, 47, 63, 66
texture, surface 47
thickness 41
problems 42
threads 3
lengths for multi-section bookbinding 156
thick linen thread 153, 156–9
tide lines 109
torn wet pulp cover 160–3, 179
turkey baster 25

varnish, yacht 30, 33
vat 22, 37, 180
adding pulp to 44
adding size to 38–9, 44
pulling a sheet of paper 40–2
vilene, iron-on 67
vinegar, white 69, 70, 71

wallpaper paste 25, 38, 62, 65–6
washfastness 180
washing line, indoor 25, 46, 47, 63, 66
wastepaper 8–9, 21
inclusions 82–5, 96–7
pre-coloured 51–2
water 36, 37, 44
watercut 108–11, 126–7, 180
waterleaf 38, 180
watermarks 112–16, 126, 127, 180
waterproof wood glue 29, 30
wave patterns 144–8, 149, 151
waxed thread 156
wet lap 9–11, 180
wetting agent 70, 180
white vinegar 69, 70, 71
wire, copper 115
wood glue, waterproof 29, 30
wooden board 114
work surface, preparing 39

yacht varnish 30, 33

TITLES AVAILABLE FROM
GMC Publications
BOOKS

WOODWORKING

Advanced Scrollsaw Projects	GMC Publications
Beginning Picture Marquetry	Lawrence Threadgold
Bird Boxes and Feeders for the Garden	Dave Mackenzie
Celtic Carved Lovespoons: 30 Patterns	Sharon Littley & Clive Griffin
Celtic Woodcraft	Glenda Bennett
Complete Woodfinishing (Revised Edition)	Ian Hosker
David Charlesworth's Furniture-Making Techniques	David Charlesworth
David Charlesworth's Furniture-Making Techniques – Volume 2	David Charlesworth
The Encyclopedia of Joint Making	Terrie Noll
Furniture-Making Projects for the Wood Craftsman	GMC Publications
Furniture-Making Techniques for the Wood Craftsman	GMC Publications
Furniture Projects with the Router	Kevin Ley
Furniture Restoration (Practical Crafts)	Kevin Jan Bonner
Furniture Restoration: A Professional at Work	John Lloyd
Furniture Restoration and Repair for Beginners	Kevin Jan Bonner
Furniture Restoration Workshop	Kevin Jan Bonner
Green Woodwork	Mike Abbott
Intarsia: 30 Patterns for the Scrollsaw	John Everett
Kevin Ley's Furniture Projects	Kevin Ley
Making Chairs and Tables	GMC Publications
Making Chairs and Tables – Volume 2	GMC Publications
Making Classic English Furniture	Paul Richardson
Making Heirloom Boxes	Peter Lloyd
Making Screw Threads in Wood	Fred Holder
Making Shaker Furniture	Barry Jackson
Making Woodwork Aids and Devices	Robert Wearing
Mastering the Router	Ron Fox
Pine Furniture Projects for the Home	Dave Mackenzie
Practical Scrollsaw Patterns	John Everett
Router Magic: Jigs, Fixtures and Tricks to Unleash your Router's Full Potential	Bill Hylton
Router Tips & Techniques	Robert Wearing
Routing: A Workshop Handbook	Anthony Bailey
Routing for Beginners	Anthony Bailey
Sharpening: The Complete Guide	Jim Kingshott
Sharpening Pocket Reference Book	Jim Kingshott
Simple Scrollsaw Projects	GMC Publications
Space-Saving Furniture Projects	Dave Mackenzie
Stickmaking: A Complete Course	Andrew Jones & Clive George
Stickmaking Handbook	Andrew Jones & Clive George
Storage Projects for the Router	GMC Publications
Test Reports: The Router and Furniture & Cabinetmaking	GMC Publications
Veneering: A Complete Course	Ian Hosker
Veneering Handbook	Ian Hosker
Woodfinishing Handbook (Practical Crafts)	Ian Hosker
Woodworking Techniques and Projects	Anthony Bailey
Woodworking with the Router: Professional Router Techniques any Woodworker can Use	Bill Hylton & Fred Matlack

UPHOLSTERY

The Upholsterer's Pocket Reference Book	David James
Upholstery: A Complete Course (Revised Edition)	David James
Upholstery Restoration	David James
Upholstery Techniques & Projects	David James
Upholstery Tips and Hints	David James

TOYMAKING

Scrollsaw Toy Projects	Ivor Carlyle
Scrollsaw Toys for All Ages	Ivor Carlyle

DOLLS' HOUSES AND MINIATURES

1/12 Scale Character Figures for the Dolls' House	James Carrington
Americana in 1/12 Scale: 50 Authentic Projects	Joanne Ogreenc & Mary Lou Santovec
Architecture for Dolls' Houses	Joyce Percival
The Authentic Georgian Dolls' House	Brian Long
A Beginners' Guide to the Dolls' House Hobby	Jean Nisbett
Celtic, Medieval and Tudor Wall Hangings in 1/12 Scale Needlepoint	Sandra Whitehead
Creating Decorative Fabrics: Projects in 1/12 Scale	Janet Storey
The Dolls' House 1/24 Scale: A Complete Introduction	Jean Nisbett
Dolls' House Accessories, Fixtures and Fittings	Andrea Barham
Dolls' House Furniture: Easy-to-Make Projects in 1/12 Scale	Freida Gray
Dolls' House Makeovers	Jean Nisbett
Dolls' House Window Treatments	Eve Harwood
Easy to Make Dolls' House Accessories	Andrea Barham
Edwardian-Style Hand-Knitted Fashion for 1/12 Scale Dolls	Yvonne Wakefield
How to Make Your Dolls' House Special: Fresh Ideas for Decorating	Beryl Armstrong
Make Your Own Dolls' House Furniture	Maurice Harper
Making 1/12 Scale Wicker Furniture for the Dolls' House	Sheila Smith
Making Dolls' House Furniture	Patricia King
Making Georgian Dolls' Houses	Derek Rowbottom
Making Miniature Chinese Rugs and Carpets	Carol Phillipson
Making Miniature Food and Market Stalls	Angie Scarr
Making Miniature Gardens	Freida Gray
Making Miniature Oriental Rugs & Carpets	Meik & Ian McNaughton
Making Period Dolls' House Accessories	Andrea Barham
Making Tudor Dolls' Houses	Derek Rowbottom
Making Victorian Dolls' House Furniture	Patricia King
Medieval and Tudor Needlecraft: Knights and Ladies in 1/12 Scale	Sandra Whitehead
Miniature Bobbin Lace	Roz Snowden
Miniature Embroidery for the Georgian Dolls' House	Pamela Warner
Miniature Embroidery for the Tudor and Stuart Dolls' House	Pamela Warner
Miniature Embroidery for the Victorian Dolls' House	Pamela Warner
Miniature Needlepoint Carpets	Janet Granger
More Miniature Oriental Rugs & Carpets	Meik & Ian McNaughton
Needlepoint 1/12 Scale: Design Collections for the Dolls' House	Felicity Price
New Ideas for Miniature Bobbin Lace	Roz Snowden
Patchwork Quilts for the Dolls' House: 20 Projects in 1/12 Scale	Sarah Williams

CRAFTS

American Patchwork Designs in Needlepoint	Melanie Tacon
Bargello: A Fresh Approach to Florentine Embroidery	Brenda Day
Beginning Picture Marquetry	Lawrence Threadgold
Blackwork: A New Approach	Brenda Day
Celtic Cross Stitch Designs	Carol Phillipson
Celtic Knotwork Designs	Sheila Sturrock
Celtic Knotwork Handbook	Sheila Sturrock

Celtic Spirals and Other Designs — Sheila Sturrock
Complete Pyrography — Stephen Poole
Creating Made-to-Measure Knitwear: A Revolutionary Approach to
 Knitwear Design — Sylvia Wynn
Creative Backstitch — Helen Hall
Creative Embroidery Techniques Using Colour Through Gold
— Daphne J. Ashby & Jackie Woolsey
The Creative Quilter: Techniques and Projects — Pauline Brown
Cross-Stitch Designs from China — Carol Phillipson
Cross-Stitch Floral Designs — Joanne Sanderson
Decoration on Fabric: A Sourcebook of Ideas — Pauline Brown
Decorative Beaded Purses — Enid Taylor
Designing and Making Cards — Glennis Gilruth
Glass Engraving Pattern Book — John Everett
Glass Painting — Emma Sedman
Handcrafted Rugs — Sandra Hardy
Hobby Ceramics: Techniques and Projects for Beginners — Patricia A. Waller
How to Arrange Flowers: A Japanese Approach to English Design
— Taeko Marvelly
How to Make First-Class Cards — Debbie Brown
An Introduction to Crewel Embroidery — Mave Glenny
Making and Using Working Drawings for Realistic Model Animals
— Basil F. Fordham
Making Character Bears — Valerie Tyler
Making Decorative Screens — Amanda Howes
Making Fabergé-Style Eggs — Denise Hopper
Making Fairies and Fantastical Creatures — Julie Sharp
Making Greetings Cards for Beginners — Pat Sutherland
Making Hand-Sewn Boxes: Techniques and Projects — Jackie Woolsey
Making Mini Cards, Gift Tags & Invitations — Glennis Gilruth
Making Soft-Bodied Dough Characters — Patricia Hughes
Native American Bead Weaving — Lynne Garner
Natural Ideas for Christmas: Fantastic Decorations to Make
— Josie Cameron-Ashcroft & Carol Cox
New Ideas for Crochet: Stylish Projects for the Home — Darsha Capaldi
Papercraft Projects for Special Occasions — Sine Chesterman
Patchwork for Beginners — Pauline Brown
Pyrography Designs — Norma Gregory
Pyrography Handbook (Practical Crafts) — Stephen Poole
Rose Windows for Quilters — Angela Besley

Rubber Stamping with Other Crafts — Lynne Garner
Silk Painting — Jill Clay
Sponge Painting — Ann Rooney
Stained Glass: Techniques and Projects — Mary Shanahan
Step-by-Step Pyrography Projects for the Solid Point Machine
— Norma Gregory
Tassel Making for Beginners — Enid Taylor
Tatting Collage — Lindsay Rogers
Tatting Patterns — Lyn Morton
Temari: A Traditional Japanese Embroidery Technique — Margaret Ludlow
Trip Around the World: 25 Patchwork, Quilting and Appliqué Projects
— Gail Lawther
Trompe l'Oeil: Techniques and Projects — Jan Lee Johnson
Tudor Treasures to Embroider — Pamela Warner
Wax Art — Hazel Marsh

PHOTOGRAPHY

Close-Up on Insects — Robert Thompson
Double Vision — Chris Weston & Nigel Hicks
An Essential Guide to Bird Photography — Steve Young
Field Guide to Bird Photography — Steve Young
Field Guide to Landscape Photography — Peter Watson
How to Photograph Pets — Nick Ridley
In my Mind's Eye: Seeing in Black and White — Charlie Waite
Life in the Wild: A Photographer's Year — Andy Rouse
Light in the Landscape: A Photographer's Year — Peter Watson
Outdoor Photography Portfolio — GMC Publications
Photographing Fungi in the Field — George McCarthy
Photography for the Naturalist — Mark Lucock
Professional Landscape and Environmental Photography:
 From 35mm to Large Format — Mark Lucock
Rangefinder — Roger Hicks & Frances Schultz
Viewpoints from Outdoor Photography — GMC Publications
Where and How to Photograph Wildlife — Peter Evans

ART TECHNIQUES

Oil Paintings from your Garden: A Guide for Beginners — Rachel Shirley

MAGAZINES

WOODTURNING ◆ WOODCARVING ◆ FURNITURE & CABINETMAKING
THE ROUTER ◆ NEW WOODWORKING ◆ THE DOLLS' HOUSE MAGAZINE
OUTDOOR PHOTOGRAPHY ◆ BLACK & WHITE PHOTOGRAPHY
TRAVEL PHOTOGRAPHY
MACHINE KNITTING NEWS ◆ BUSINESSMATTERS

The above represents a selection of the titles currently published or scheduled to be published.
All are available direct from the Publishers or through bookshops, newsagents and specialist retailers.
To place an order, or to obtain a complete catalogue, contact:

GMC Publications,
Castle Place, 166 High Street, Lewes, East Sussex BN7 1XU, United Kingdom
Tel: 01273 488005 Fax: 01273 402866
E-mail: pubs@thegmcgroup.com

Orders by credit card are accepted